The *Natural History* Reader in Animal Behavior

THE *NATURAL HISTORY* READER
in ANIMAL BEHAVIOR

HOWARD TOPOFF, *Editor*

New York   COLUMBIA UNIVERSITY PRESS   *1987*

Library of Congress Cataloging-in-Publication Data

The Natural history reader in animal behavior.

Includes index.
1. Animal behavior.   I. Topoff, Howard R.
II. Natural history.
QL751.6.N38   1987   591.5'1   86-26848
ISBN 0-231-06158-7
ISBN 0-231-06159-5 (pbk.)

Columbia University Press
New York   Guildford, Surrey
Copyright © 1987 Columbia University Press
All rights reserved

Printed in the United States of America

Clothbound editions of Columbia University Press books
are Smyth-sewn and printed on permanent and durable acid-free paper.

Contents

Original Publication Dates of Articles   vii

Introduction   1

Additional Readings   15

PART 1. SENSORY PROCESSES AND ORIENTATION   17

1. A Special Light to Steer By
    WILLIAM G. WELLINGTON   19

2. The Shark's Sixth Sense
    ADRIANUS J. KALMIJN AND KENNETH JON ROSE   28

3. Night Fighters in a Sonic Duel
    KENNETH D. ROEDER   34

4. The Pit and the Antlion   HOWARD TOPOFF   44

5. Bird Navigation:
    Travels Around New England in Pursuit of Pigeons
    CHARLES WALCOTT   55

6. Psychophysics and Hearing in Fish
    WILLIAM N. TAVOLGA   66

7. Invertebrate Learning   MARTIN J. WELLS   75

PART 2. EVOLUTION AND BEHAVIOR   85

8. Evolution of Nest Building
    NICHOLAS E. COLLIAS   87

9. Evolution of the Web   B.J. KASTON   97

10. The Importance of Being Feverish
    MATTHEW J. KLUGER   106

11. Goose Mates   FRED COOKE   113

12. Strategies of Reproduction
    R. D. MARTIN   120

PART 3. SOCIAL ORGANIZATION   127

13. Four Months of the Ground Squirrel
    PAUL W. SHERMAN AND
    MARTIN L. MORTON   129

14. The Helpful Shall Inherit the Scrub
    JOHN W. FITZPATRICK AND
    GLEN E. WOOLFENDEN   138

15. The Hummingbird and the Calorie
    PAUL W. EWALD   147

16. Invasion of the Booty Snatchers
    HOWARD TOPOFF   155

17. Fish in Schools   EVELYN SHAW   162

18. Masters of the Tongue Flick
    CAROL A. SIMON   168

19. Peaceable Peccaries   JOHN A. BYERS   177

20. New Theory on Fabled Exodus
    KAI CURRY-LINDAHL   186

PART 4. BEHAVIORAL DEVELOPMENT   195

21. Components of Recognition in Ducklings
    GILBERT GOTTLIEB   197

22. Mother Baboon
    JOAN LUFT AND JEANNE ALTMANN   203

23. A Mound of One's Own
    W. THOMAS JONES AND BETSY BUSH   215

24. Sea Lion Shenanigans
    FRED BRUEMMER   221

25. At Play in the Fields
    PHYLLIS DOLHINOW   229

Index   239

Original Publication Dates of Articles
in *Natural History* Magazine

Sea Lion Shenanigans
by Fred Bruemer   (1983), 7:32–41

Peaceable Peccaries
by John A. Byers   (1981), 90(6):60–66

Evolution of Nest Building
by Nicholas E. Collias   (1965), 74(7):40–47

Goose Mates
by Fred Cooke   (1983), 92(1):36–43

New Theory on a Fabled Exodus
by Kai Curry-Lindahl   (1963), 72(7):46–53

At Play in the Fields
by Phyllis Dalhinow   (1971), 80(10):66–71

The Hummingbird and the Calorie
by Paul W. Ewald   (1979), 88(7):92–98

The Helpful Shall Inherit the Scrub
by John W. Fitzpatrick and Glen E. Woolfenden   (1984), 93(5):55–63

Components of Recognition in Ducklings
by Gilbert Gottlieb   (1965), 74(2):12–19

A Mound of One's Own
by Thomas W. Jones and Betsy Bush   (1984), 93(11):60–67

The Shark's Sixth Sense
by Adrianus Kalmign and Kenneth Jon Rose   (1978), 87(3):76–81

Evolution of the Web
by B. J. Kaston   (1966), 75(4):26–33

The Importance of Being Feverish
by Matthew J. Kluger   (1976), 85(1):70–75

Mother Baboon
  by Joan Luft and Jeanne Altmann   (1982), 91(9):30–39

Strategies of Reproduction
  by R. D. Martin   (1975), 84(9):48–57

Night Fighters in a Sonic Duel
  by Kenneth D. Roeder   (1964), 73(1):32–39

Fish in Schools
  by Evelyn Shaw   (1975), 84(8):40–46

Four Months of the Ground Squirrel
  by Paul W. Sherman and Martin L. Morton   (1979), 88(6):50–57

Masters of the Tongue Flick
  by Carol A. Simon   (1982), 91(9):58–67

Psychophysics and Hearing in Fish
  by William N. Tavolga   (1964), 73(3):34–41

The Pit and the Antlion
  by Howard Topoff   (1977), 86(4):64–71

Invasion of the Booty Snatchers
  by Howard Topoff   (1984), 93(10):78–85

Bird Navigation
  by Charles Walcott   (1972), 81(6):32–43

A Special Light to Steer By
  by William G. Wellington   (1974), 83(10):46–53

Invertebrate Learning
  by Martin J. Wells   (1966), 75(2):34–41

The *Natural History* Reader in Animal Behavior

Introduction

The modern study of comparative animal behavior represents a marriage, and like all such unions, it has incorporated something borrowed, something new, and something old (as far as something blue is concerned, well no marriage is perfect). What's borrowed is that the scientific study of animal behavior was made possible by the synthesis of principles and techniques from two well-established fields: zoology and psychology. What's new is that, as an objective discipline, animal behavior is one of the youngest branches of the behavioral sciences. But paradoxically, and this is what's old, observations of animal behavior undoubtedly predate the evolutionary origin of human beings. After all, even our nearest nonhuman primate relatives take advantage of their keen observational abilities to prey upon other animal species, as illustrated by the following description (by Jane van Lawick-Goodall) of predation by chimpanzees.

> A group of chimps rested in the shade of a tall tree. In its branches, a juvenile baboon fed alone, separated by some 200 yards from the rest of his troop. Presently Huxley (a mature male) plodded up from the stream toward the peaceful chimpanzee group. About 10 feet from the fig tree he stopped, facing its trunk. To us he seemed unaware even of the existence of the small baboon above. Nonetheless as though he had in fact given a signal, the other chimps stood up. Two of the males moved to the base of the fig tree; three others stationed themselves under two nearby trees, the branches of which formed an escape route for the baboon. And then, very slowly, with infinite caution, Figan, the youngest of the males present (he was about eight at the time), began to creep toward his quarry.

In prehistoric times, humans were probably able to subsist on a variety of plant foods, much as many societies do today. But the fossil records of tools and carcasses clearly show that, like other primates, our earliest protohuman ancestors in Africa supplemented their diet with the meat of animals that were scavenged or killed. Indeed, the availability of clothing as well as food was undoubtedly determined by the success of the hunt. Now chimpanzee hunting behavior is not notable for any significant degree of foresight. Instead of seeking out their prey, apes typically take advantage of animals that wander into vulnerable areas. But just imagine how much more efficient predation became when human ancestors perfected the ability to determine when and where animals migrated during different seasons, where animals went for food and water, for courtship and mating, or even to sleep. In short, successful hunters and fishermen were excellent animal behaviorists!

Why Study Animal Behavior?

In today's technological world, our interest in animal behavior can still be quite utilitarian. For example, studies of orientation and habitat selection were crucial for understanding and controlling the periodic outbreaks of bubonic plague, typhus, yellow fever, malaria, and other diseases spread by arthropod vectors. And of course the success of our agricultural industries depends upon knowledge of the feeding and reproductive behavior of livestock and poultry, as well as the behavior of insect pests. In the case of one well-known pest, the screw worm fly, behavioral research led to an almost complete solution to what has been for many years an intractable agricultural problem. The screw worm is a fly that lays its eggs in the wounds of cattle. The hundreds of eggs in each sore hatch into maggots (larvae) that feed on the cow's flesh, debilitating and sometimes killing infested animals. Successful control by the United States Department of Agriculture was made possible by research which showed that the female fly has a rather deprived sex life: she mates only once in her lifetime. So over a period of years, billions of male flies were sterilized by radiation (which does not alter their mating behavior) and released from airplanes, free to compete with wild fertile males for females. As a result of this "flies from the skies" program, more females mated with the irradiated males than with normal flies, and those that did laid only sterile eggs. Thus the screw worm fly was virtually eliminated from large areas of the United States.

A second practical reason for studying animal behavior is to increase our expertise in animal conservation—in zoos, preserves, and national parks. One of the most publicized examples of this kind of research stemmed from campers in Yellowstone National Park who had been injured or killed by grizzly bears. Although the simplest solution to this problem would be to eliminate the bears altogether, most people would agree that national parks exist in part to maintain and protect the animals, especially endangered species. Behavioral and ecological studies sponsored by the National Park Service showed that the movements of some grizzly bears in the Park were influenced predominantly by the location of garbage dumps. In a completely natural setting, bears would undoubtedly avoid human beings. But in Yellowstone, frequent contact with garbage has conditioned the bears to associate human odors with food. To reverse this trend, the Park Service is gradually removing the garbage dumps to more remote areas, thereby encouraging bears to stay away from the campgrounds. In the terminology of learning theory, the reinforcer (food) is being removed and the conditioned response of the bears extinguished.

For many people, the essential function of science (and the justification for its support by society) is to improve the quality of life by increasing our control over the environment. As the examples discussed above show, animal behavior shares this important goal with other sciences. But there is a second and equally important function that is often taken for granted; in addition to a technological role, the behavioral scientist's activities have much in common with those of artists.

Through their creations, artists share with us their private, very personal views of the world, and we support their accomplishments primarily because we find our lives enriched by the unique perspectives provided by their work. Now consider that in the real world there are well over one million species of living animals, and that each has evolved unique behavioral adaptations for feeding, fighting, mating, and finding a place to live. These patterns of behavior, which include (just to name a few) female black widow spiders consuming their mates after copulation, birds using the earth's geomagnetic field for navigation, and ants capturing the young of foreign species to raise as slaves, are every bit as fascinating, bizarre, and mysterious as any artistic creation. Unfortunately, most of us are not privy to these activities, even though they routinely take place in the natural world that surrounds us. But the scientist, who is specifically trained in the use of both experimental methodology and equipment, is precisely the person who uncovers these secrets and shares them with us. Thus through science, as in art, we capture the richness of the world and not just its riches.

Truth in Labeling

The development of animal behavior during the past 40 years into an objective, experimental science was made possible only by the gradual elimination of our anthropomorphic bias, a conceptual obstacle that seems to haunt the natural sciences especially. In physics and chemistry, for example, we do not typically explain the mutual approach of positively and negatively charged particles by attributing human feelings to them. And while it's true that such anthropomorphic labels as "attraction" are often used to describe these movements, no one seriously entertains the notion that an electron is sad when it is not near a proton. But in the study of animal behavior, no doubt because we humans are part of the animal kingdom, three forms of anthropomorphism do persistently crop up. First, the terms used to describe the activities of animals often become confused with interpretations about the causes of behavior. For example, if a social wasp is observed at her nest entrance attempting to sting an intruding individual from another colony, we would probably use the word "aggression" to characterize her behavior. But we must remember that "aggression" is only a descriptive term, a label that humans assign to patterns of behavior in other species if their behavior looks in any way similar to human fighting. And the same holds for virtually all other descriptive terms, such as feeding, courtship, play, or even sleep. The point is that each of these labels represents an equivalent adaptive outcome, which enables numerous species of animals to solve common ecological problems. But using the same anthropomorphic label to describe the behavior of organisms as diverse as wasps and humans can be misleading because the mechanisms under-lying these behaviors may be qualitatively different for the two animal groups. Indeed, this is precisely why "comparative animal behavior" is defined as the scientific study of the similarities and differences in the mechanisms underlying the behavior of species representing all levels of evolutionary history.

A second (although related) type of anthropomorphism concerns the equally common tendency to assume that animals may be described as if they have human feelings, needs, and abilities. As an example, the Swiss zoologist H. Hediger once reported on how a gate in a zoo was inadvertently left open, enabling a herd of roe deer to leave their enclosure and disappear into a nearby forest. Although the deer could have easily survived in the wild, surprisingly they promptly returned to the zoo. By analogy to humans, it was as though prisoners had returned to their place of detention.

Although Hediger did not conduct any in-depth studies of the deer, we can easily propose an objective and nonanthropomorphic hypothesis to explain their apparently odd behavior. For instance, many mammals feed and mate in a well-defined area called a "home range." The movement of the deer into the forest may have been an example of their tendency to explore novel environments, and their return to the enclosure may, in turn, reflect their prior attachment to an area that provided adequate food, shelter, and the absence of predators. In other words, what to humans looks like a "prison" might, to the deer, simply represent the core area of their home range!

The last example of anthropomorphism is evidenced by our tendency to rank the behavior of animals on some kind of linear scale of intelligence, once again using humans as the standard for comparison. Surprisingly, much of the blame for this fallacy can be attributed to, of all people, Charles Darwin. It is generally agreed that by the time *On the Origin of Species* was published in 1859, Darwin had alread y gathered much of the evidence needed to incorporate humans into his evolutionary framework. Many historians maintain that Darwin judiciously avoided the question of human origins and behavior in *Origin,* because he knew he was in for a hard enough time even without this more controversial topic. In any case, he waited 12 years to publish *The Descent of Man,* the book which launched the study of comparative animal behavior. As is well known, Darwin's principal thesis was that the resemblances between humans and other primates constituted clear evidence of gradual descent, with modification, from a common ancestor. And because of these anatomical similarities, Darwin claimed that the behavior of humans and other animal species also differs not qualitatively, but only in degree. Our level of intelligence, he maintained, can be thought of merely as a point on a behavioral continuum that runs through all animal species.

By emphasis on this principle of behavioral continuity, humans became the standard by which all other species were judged. And because the outstanding characteristic of humans is their ability to solve complex problems, a generation of psychologists converged on a common research program in which animals representing many evolutionary lineages were tested in mazes, puzzle boxes, and a host of other contraptions that simulated human-oriented tasks. Then, on the basis of their performance, the various species were ranked on a scale of intelligence (the top of which was, of course, one notch below the spot preassigned to humans).

The problem with this concept of intelligence is that achieving human status is not the goal of evolution. Each species has a unique evolutionary history and may therefore be adapted for surviving in qualitatively different ecological niches. As a result, the behavioral capabilities of animals should be judged within the context of their ecological setting. Once again, an example will illustrate this point.

The Nose Knows

Many species of salmon in western North America hatch in small mountain streams and migrate down river to the Pacific Ocean. After several years of marine life, the adult salmon return to their freshwater stream of origin and reproduce. In one very ambitious research project, almost 500,000 newly hatched sockeye salmon were tagged in a small tributary of the Fraser River in western Canada.

Several years later, after the fishes' return migration, nearly 11,000 individuals were recovered in the same tributary. Surprisingly, not one of the tagged salmon was ever found in a different stream. How did the salmon select the correct junction at each of the numerous tributary forks encountered on their upstream run? Arthur Hasler and his colleagues at the Wisconsin Lake Laboratory tackled this problem by testing the hypothesis that the fishes orient themselves by responding to chemicals in the water. In one experiment they used sexually mature salmon that had recently returned to each of two branches of the Issaquah River in Washington. Using cotton, they plugged the noses of half the fish in each stream group, and then placed all the salmon back in the river below the fork leading to the two branches. As predicted, most of the fishes with unplugged noses swam back to the stream they had originally selected. The experimental group, by contrast, migrated randomly, with roughly equal numbers swimming to both branches. In subsequent tests, the researchers were able to demonstrate that tributaries differ with respect to the chemical properties of their waters, and that the adult salmon can remember the odors to which they were exposed on their downstream journey shortly after hatching.

Now, by way of comparison, suppose we construct a "Y"-shaped tunnel large enough to accommodate a human being, and suspend the tunnel five feet under water in a swimming pool. A human subject is placed at the base of the "Y" and swims upstream toward the common junction of the three runways. We then introduce a chemical into the water, flowing through one of the two upper tunnels. Could the human subject (the most "intelligent" of all species), like the salmon, discriminate between the two chambers on the basis of the chemical difference of the water? The answer is simple: no way! The human nose evolved in a terrestrial environment, and is virtually useless in an aquatic medium.

Now suppose a group of salmon decided to rank all other species of animals on some scale of intelligence. Undoubtedly, after a little deliberation, they would select a behavioral task that was central to their own evolutionary success. And since chemoreception under water, in addition to facilitating migration, also plays a vital role in feeding, courtship, and avoiding predators, an aquatic learning problem would be a logical choice for their (salmonpomorphic) intelligence test. Now we can be sure that the particular task selected would result in the salmon scoring the highest marks on such a test. Next would probably come other fishes, followed perhaps by related aquatic vertebrates, and so on. Humans, unfortunately, would flunk miserably, and be placed way down on the salmon's evolutionary scale of IQ.

This example is not meant to suggest that meaningful comparative studies of learning are impossible to conduct. The point is that the use of humans as the standard of comparison is an anthropomorphic procedure that ignores the evolutionary background of each species, and the unique adaptations that it may possess for surviving and reproducing in its environment.

Levels of Integration in the Study of Behavior

Because animal behavior is a relatively young science, perhaps much can be learned from other (and more established) disciplines within the biological sciences. For example, comparative anatomists have for more than 100 years been examining the similarities and differences in animal morphology. Together with paleontologists, they have used the results of their studies to establish evolutionary relationships among these species. It might therefore seem that a great deal of the behaviorists' job had been completed. After all shouldn't those species that are most similar morphologically also be most similar behaviorally? The answer, surprisingly, is a definite maybe! To see why this is so, let's explore the concept of "levels of integration," which was developed for animal behavior by the late T. C. Schneirla, and which provides an important theoretical basis for comparative studies. Although it has a rather ominous sounding name, the levels concept is a sort of reformulation of a very familiar idea—namely, that the whole is greater than the sum of its parts. Accordingly, the biotic world is viewed as being organized into a series of levels of organization of increasing complexity. Each level exhibits unique properties that appear only when the units it comprises interact. The key word here is "interact," whose meaning can be easily illustrated with an example from chemistry. Suppose you are given two flasks, one containing pure oxygen and the other hydrogen. You may conduct any measurements on these two gases, so long as you don't combine them. From these calculations, could you specify the freezing point, boiling point, and density of water? The answer is obviously no, because when oxygen and hydrogen interact, the product (water, in this case) exhibits emergent properties that are qualitatively different from those of its individual components. In other words, if you want to elucidate the "behavior" of water, you have no choice but to conduct an empirical investigation at the appropriate level: simply put, you have to study water.

Now the behavior of animals is also an emergent level of organization, because behavior is influenced by an interaction among its components. These include (to name just a few) genetic processes, morphology, endocrine secretions, and the physiological functioning of all additional organ systems. As a result, relatively small, quantitative changes in any one characteristic during the course of evolution could produce gross, qualitative changes in behavior. As an example, consider that human beings are classified in the primate family Hominidae, while the great apes are placed in a separate family, the Pongidae. Although the separation of species into different families usually implies a rather wide gap in biological characteristics, there do not seem to be sufficient morphological differences between these groups to warrant such divergent taxonomic placement. Indeed, recent biochemical studies comparing the proteins and nucleic acids among primates show that humans and chimpanzees share about the same percentage of genes as do sibling species (belonging to one genus) of the fruit fly *Drosophila*. During the course of primate evolution, mutations have occurred in a small number of what are termed "switch genes," any one of which may influence numerous developmental processes. As a result, the phenotypic outcome of these relatively minor genetic substitutions for *Homo sapiens* were qualitative alterations in the organization of the central nervous system and in the behavioral capabilities which it produced.

In most taxonomic studies, decisions concerning evolutionary similarity are based on analyses of a limited number of traits. Because behavior is influenced by interactions among characteristics at lower levels of organization, and because these characteristics often evolve in a mosaic fashion (i.e., at different rates), we cannot automatically conclude that the psychological capabilities of two species are as similar as biological characteristics operating at lower levels of organization. Thus, the heart of a pig may be an excellent model for a human heart, but the mechanisms underlying fighting behavior in pigs may shed little light on human aggression. The nervous systems of both humans and earthworms contain opiate-like chemicals called endorphins, which are secreted during periods of intense arousal, but this neurophysiological fact alone does not tell us whether annelid worms experience stress or pain in the human sense. And finally, about 10 percent of human males report sexual impotency after a vasectomy. That this problem is psychological (e.g., due to a castration complex), not physiological, is suggested by the fact that other primate species experience no change in sexual behavior after the same surgical procedure. On the anatomical and physiological levels of organization, humans and other primates have remarkably similar reproductive systems. But on the behavioral level, only humans are capable of reacting to a vasectomy with a degree of psychological complexity that produces sexual impotency.

The moral of the "levels" concept is therefore simple: The study of animal behavior must be based on empirical analyses of the behavior of numerous invertebrate and vertebrate species. If we simply sit back and make assumptions about behavioral differences among species solely on the basis of the analyses conducted at lower levels of organization, we run the risk of putting the theoretical cart before the factual horse.

The Four Questions About Animal Behavior

Once the empirical foundation is provided through detailed observations and descriptions of animal behavior, the investigator may then proceed to ask a variety of specialized questions, most of which fall into four categories: (1) immediate causation; (2) development; (3) adaptive function; and (4) evolution. The distinctions among these four categories can best be illustrated by applying them to one of my own research projects—the social organization of army ants.

The designation "army ants" subsumes over 300 species with rather diverse ecological and behavioral characteristics. The geographical range of most species is restricted to the tropics, where we find the largest colonies. In Central and South America, for example, some species of the genus *Eciton* have colonies containing well over 250,000 individuals. And in Africa, estimates of colony size for the genus *Dorylus* range up to 50 million.

Army ants exhibit an assemblage of behavioral features that sets them apart from most other ants. Perhaps foremost among these is their periodic emigrations to new nesting sites. These journeys are dramatic events in which the adult worker ants transport the entire brood population to a new nesting site, which can be several hundred meters away. Moreover, in genera such as *Eciton* and *Neivamyrmex,* these emigrations occur with remarkable regularity, as part of a behavioral cycle of alternating nomadic and stationary phases. Pioneering studies conducted by Schneirla showed that during the nomadic phase, which lasts about two and one-half weeks, the large predatory raids conducted each day are followed by nighttime emigrations to new bivouacs. During the three-week long stationary phase, by contrast, the intensity and duration of raiding is reduced, and the colony remains at the same nesting site.

In addition to emigrations, army ants are notorious for their ability to conduct large-scale raids of impressive magnitude. Indeed, it is their mastery of group predation that gave army ants the nickname "Huns and Tartars" of the insect world. In many other types of ants, workers forage individually and recruit nestmates only after locating a source of food. This behavioral strategy succeeds when the colony's nutritional requirements are satisfied by nectar, a carcass, or any other kind of immobile food. But for a predator such as army ants, which specializes on mobile social insects, raiding success requires that a large and replenishable striking force be mobilized at the target colony within seconds after its discovery. Army ants meet this ecological challenge by conducting all raids in well-organized groups containing tens of thousands of highly aroused individuals. In the neotropical species *Eciton hamatum* and in *Neivamyrmex nigrescens* from the United States, a single basal ant column extends from the nest at the very onset of raiding, and then splits into a dendritic network of trails. In other species, including the related *E. burchelli* and the genus *Labidus,* the basal column typically terminates in a fan-shaped swarm that is several meters wide at its distal end.

(1) Immediate causation

Here the scientist is interested in the processes that influence behavior in the short term. These include the responses of sensory organs to environmental stimuli, neural functioning, the role of hormones and pheromones, and all other physio-logical systems that regulate behavior at the moment of its occurrence. For example, the key to efficient group predation by army ants is their use of a communication process known as "chemical mass recruitment." When army ants advance along the terrain in search of food, each worker deposits a chemical trail from its hindgut. This exploratory trail is relatively stable, and serves as an orienting stimulus for nestmates to follow. During this exploratory phase, traffic along the trail is bidi-rectional, with approximately equal numbers of workers moving to and from the nest. When a suitable prey colony of termites, wasps, or ants is discovered, army ants at the raiding front become highly aroused and begin recruiting nestmates to the target. The recruiters run back and forth along a small segment of the trail, contacting all other ants encountered in their path. Field studies on *Eciton* and *Neivamyrmex* showed that these recruiting ants deposit a pheromone that is different from the exploratory trail. This recruitment pheromone is considerably more volatile than the exploratory trail, and it is unique in its ability to generate secondary recruiters. This means that if an army ant worker contacts the trail deposited by an excited recruiter, that worker behaves as if it too had found the food. Because many hundreds of army ants can become secondary recruiters, the wave of excitement that originates at the prey's nest "flows" back toward the bivouac and outward along all peripheral columns. The net result of this dynamic communication process is the arrival of thousands of biting and stinging army ants at the target site, within minutes after its initial discovery.

(2) Development

When asking questions about the development of behavior, the researcher focuses on the changes that occur during the lifespan of an individual organism. Individuals of many animal species exhibit what is called "temporal polyethism," which means that their behavioral role changes as a function of both their maturation and their experiences with environmental stimuli. Developmental (or ontogenetic) questions may span the full range of an organism's life, from conception to death. Among army ants, the young workers (called callows) that have recently emerged from the pupal stage do not participate in the predatory raids until they are about one week old. When *raiding* occurs during this period, the callows remain in the bivouac and cluster tightly together in a compact mass. This sort of temporal change in the callows' foraging behavior is very much like that found in many other social insects. But army ants are nomadic, and—to make matters worse— the first *emigration* of each new nomadic phase takes place within one day after the callows' eclosion. For the colony to remain intact, the callows must leave the nest and run the entire length of the emigration route along with the mature adult workers, after which they remain clustered in their new bivouac—until their next emigration.

A series of laboratory tests confirmed that even one-day-old callows are extremely sensitive to the chemical trail deposited along the emigration route. But during the first week after eclosion they are even more responsive to pheromone odors from the brood and from each other. During the predatory raids, adult workers recruit each other. But when a new suitable nesting site has been located at the start of an emigration, the adults use a combination of tactile and chemical stimuli to break up the mass of sedentary callows. Once sufficiently aroused, the callows leave the nest and join in the exodus to the new site.

(3) Evolution

Questions about evolution are concerned with how existing patterns of behavior arose and were modified by natural selection during the course of evolutionary history. As such, evolutionary questions relate not to particular individuals, but to behavioral changes in species. In constructing the evolutionary steps leading to army ant behavior, it is hypothesized that group raiding arose initially in relatively small colonies, as an adaptation for feeding on other social insects. This is the condition currently found in several species of ants from the primitive subfamilies Ponerinae and Cerapachyinae. Because periodic nest emigrations also occur in several ponerine species, nomadism in some species might have evolved concurrently with group predation. Later, as mass chemical recruitment became more efficient (thus enabling colonies to secure increased amounts of food), substantially larger colony sizes characteristic of many present-day army ants became possible.

(4) Function

Functional questions deal with the role that behavior plays in enabling organisms to adapt to their physical and biotic environments. Here the scientist is interested in elucidating how patterns of behavior contribute to the survival and reproductive success of the individual, and of the social group to which it belongs. Thus, to establish the function of nomadism in army ants, we start with the fact that the nomadic and stationary phases of many species are correlated with the breeding cycle within the colony. Specifically, the nomadic phase, during which the raiders bring back enormous amounts of booty, coincides with the period of larval development. Because up to 75 percent of the captured food is fed to the rapidly growing larvae, nomadic emigrations are adaptive in that they provide army ants with new areas in which to raid. When the larvae cease feeding and pupate, the stationary phase begins. During this phase, which lasts throughout pupal development, raiding is reduced in frequency and intensity, and the colony remains at the same nesting site. Since pupae require no food, the adult workers are able to capture enough for their own consumption simply by conducting a few raids from the single stationary bivouac.

The Organization of This Book

During the first part of this century, many articles in *Natural History* magazine were contributed by top-notch writers, but not necessarily the individuals who conducted the basic research. During the past 25 years, however, this emphasis has changed, so that now virtually all research reports are written by the projects' principal investigators. The advantage of this approach is that the researcher is in the best position to convey the historical origin of a scientific problem, the dedication and perseverance essential for completing the study, and of course the excitement of discovering new facts about animal behavior.

The 25 contributions selected for this book are divided into four sections: (1) Sensory Processes and Orientation; (2) Evolution and Behavior; (3) Social Organization; and (4) Behavioral Development. Needless to say, there is often a great deal of overlap between these sections, as when studies of development are aimed at elucidating the origins of social bonds, or when analyses of animal societies provide insight into evolutionary processes. My overriding concern was that the articles reflect a balance of both animal groups and behavioral processes. Although most of the articles were published during the last 10 years, I have incorporated material dating back to 1964. Even though some of these older studies may be a bit out of date by now, they nevertheless contribute to the diversity of behavioral topics that is so essential for an introductory Reader.

Howard Topoff
Department of Psychology
Hunter College of CUNY
Department of Entomology
American Museum of Natural History

Additional Readings

Alcock, J. 1984. *Animal Behavior: An Evolutionary Approach.* Sunderland, Mass: Sinauer Associates.

Barnett, S. A. 1981. *Modern Ethology: The Science of Animal Behavior.* New York: Oxford University Press.

Bermant, G., ed. 1973. *Perspectives on Animal Behavior.* Glenview, Ill.: Scott-Foresman.

Broom, D. M. 1981. *Biology of Animal Behavior.* Cambridge: Cambridge University Press.

Brown, J. L. 1975. *The Evolution of Behavior.* New York: W.W. Norton.

Dewsbury, D. A. 1978. *Comparative Animal Behavior.* New York: McGraw-Hill.

Drickamer, L. C., and S. H. Vessey. 1982. *Animal Behavior: Concepts, Processes, and Methods.* Boston: Willard Grant Press.

Eisner, T., and E. O. Wilson, eds. 1975. *Animal Behavior.* San Francisco: W.H. Freeman.

Gould, J. 1982. *Ethology: The Mechanisms and Evolution of Behavior.* New York: . Norton.

Greenberg, G., and E. Tobach, eds. 1984. *Behavioral Evolution and Integrative Levels.* Hillsdale, N.J.: Lawrence Erlbaum.

Grier, J. W. 1984. *Biology of Animal Behavior.* St. Louis: C.V. Mosby.

Hinde, R. A. 1970. *Animal Behavior: A Synthesis of Ethology and Comparative Psychology.* New York: McGraw-Hill.

Klopfer, P. H. 1974. *An Introduction to Animal Behavior: Ethology's First Century.* Englewood Cliffs, N.J.: Prentice-Hall.

Robinson, D. N., and W. R. Uttal. 1983. *Foundations of Psychobiology.* New York: Macmillan.

Topoff, H. ed. 1981. *Animal Societies and Evolution.* New York: W.H. Freeman.

PART 1
SENSORY PROCESSES AND ORIENTATION

The study of sensation is a logical topic for beginning our exploration into the world of animal behavior, because it is through their sense organs that animals receive information about their physical and social environment. Perhaps the most vivid illustration of the role of peripheral stimuli in initiating behavior is provided by the predatory and sexual behavior of the parasitic wasp *Habrobracon*. During mating the male is the more active partner, engaging in an elaborate ritual of courtship that terminates in copulation. The female, by contrast, is relatively passive throughout this period. But this lack of enthusiasm over sex by *Habrobracon* females is in marked contrast to the excitement she exhibits in the presence of a caterpillar of the moth *Ephestia*. She approaches and stings the larva, which paralyzes the prey without killing it. The female wasp then deposits her eggs on the surface of the immobile caterpillar. Upon hatching out of the eggs, the *Habrobracon* larvae live and grow by feeding upon the host. Since males of *Habrobracon* do not lay eggs, they exhibit virtually no response to the presence of *Ephestia*.

In the laboratory, it is possible to occasionally produce *Habrobracon* gynandromorphs. These are genetic intersexes in which different parts of the body have male or female organs. Now which patterns of behavior will be exhibited by an individual possessing a female abdomen and a male head? Since the female gonads, copulatory organs, and stinging apparatus are located in the abdomen, will this gynandromorph remain passive during courtship, but attempt to sting an *Ephestia* moth larva? The answer is no, because the head of the wasp is composed of "male" tissues and it is the head that contains the sensory organs for vision (eyes) and olfaction (antennae) through which the wasp perceives and reacts to its world. This type of gynandromorph has no interest in stinging prey (although anatomically it is capable of doing so), but goes wild in the presence of female wasps (even though copulation is impossible).

The following seven papers illustrate studies of sensory processes and orientation conducted on a wide variety of invertebrate and vertebrate species. The studies by Wellington on the use of polarized light by insects, and by Kalmijn and Rose on electroreception in sharks serve in part to remind us that the sensory world of other animals frequently extends well beyond the range of human sensitivity. The articles by Roeder, Topoff, and Walcott emphasize the role of sensation in specific behavioral adaptations, such as predator-prey interactions and animal migration. The paper by Tavolga on hearing in fish plays an additional role, by illustrating how techniques of operant conditioning are used to "ask" nonhuman animals about the kinds of environmental stimuli to which they can respond. Finally, the study by Wells on learning in the octopus shows how the proper choice of a sensory modality can uncover a degree of behavioral plasticity that far exceeds many people's expectations for an invertebrate species.

1

A Special Light to Steer By

WILLIAM G. WELLINGTON

When an insect flits from one plant to the next, it uses a number of directional cues. Gradients of temperature, moisture, and light; patterns of brightness and color; directional movements of air and odors—all can be used for guidance.

Most of those guides are worthless, however, when the same insect flies straight across a large open space. Neither temperature nor moisture gradients can be used to set a course across a broad lawn or a large clearing. Nor can the insect rely on wind or odors, both of which are useless for straight-line travel in the midst of an open area.

In such circumstances, visual cues become especially important. Bees and wasps, as is well known, can use landmarks in a familiar locality. But they, and many other kinds of insects, can also fly straight across featureless terrain where there are no landmarks. From time to time, therefore, scientists have suggested other visual mechanisms to account for that ability.

The most plausible—and certainly the most appealing—suggestion proposes that the insect is able to keep a straight course by continually adjusting its line of travel to hold the image of the sun steady on one part of its eye. In this "sun compass" reaction, the insect supposedly uses the image of the sun much as we use the needle of a magnetic compass.

There is no doubt that insects use the sun as a reference point. But there is mounting evidence that they do so by using the heat, not its image. Since their ability to detect temperature differences across their heat receptors is adversely affected by air currents, their "heat compass" is often least effective where it could be most useful, that is, in the middle of an open space. In fact, the most effective method available to insects for setting courses across open areas is derived, not from their capacity to sense the sun's position, but from their ability to perceive polarized light.

As early as 1940, I. Verkhovskaya discovered that fruit flies could perceive polarized light. Unfortunately, her report in the Russian literature was overlooked in the wartime disruption of international scientific communication. Then in 1948 Karl von Frisch brilliantly demonstrated that honeybees, through their dances at the hive, used polarized sky light to communicate the direction of a food source. In the wave of enthusiasm that followed his demonstration, many other investigators soon found that several kinds of animals, including insects, not only perceived polarized light but also used it in directed movements.

Because the experimental method usually required observations of the animals' responses under polarizing filters, those initial studies had to be conducted with specimens that were prevented from flying. Consequently, except for honeybees, which could be observed arriving at, and departing from, a site, no flying insect was intensively studied in its natural setting and not even honeybees could be tested in free flight.

Before that gap could be filled, interest shifted from the behavioral consequences of polarized-light perception to its physiological and structural bases. Since our own capacity to perceive polarized light is limited, the urge to discover how insects can see so clearly what we can only imagine is understandable. But hot pursuit of that physiological "how?" led to neglect of the behavioral "what then?" and thus to continued ignorance of the ecological consequences. No one knew, for example, whether an insect's ability to perceive polarized light affected aspects of its daily life other than travel. Indeed, no one asked. So some of our more unrealistic views about an insect's world remained intact longer than they should have. Before we look into that world from the insect's point of view, however, we should consider the phenomenon that has so effectively obscured our perception of it: the polarization of sky light.

A wave of sunlight approaching the atmosphere vibrates in virtually every plant at right angles to its direction of travel, that is, it is unpolarized. After entering the atmosphere, however, the wave may collide frequently with gas molecules or minute dust particles. During these collisions, some planes of vibration will be eliminated. Finally, the traveling wave may vibrate in only one plane; it has been plane polarized.

The percentage of polarized light varies in different parts of the sky (figure 1.1). There is none immediately around the sun or at the point directly opposite. The greatest amount is always located midway between these two points, at a 90-degree angle to the line joining them. The geometry is most easily visualized near sunrise and sunset. At those times the solar and antisolar points are both close to their respective horizons, and the patch of most intense polarization in

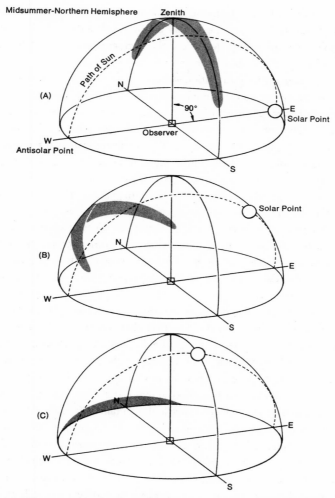

Figure 1.1. The amount of polarized light varies in intensity in different parts of the sky. The area of greatest intensity (dark grey patch) occurs midway between the solar and antisolar points. The amount of polarization decreases as the observer looks toward these points. *At sunrise (a), when the sun is barely above the horizon, the area of maximum polarization is in the zenith, directly above the observer's head. At midmorning (b), as the sun rises higher in the sky, its antisolar point sinks, and the area of greatest polarization shifts below the zenith toward the northwestern part of the sky. By noon (c), the sun is nearly overhead and the area of greatest intensity has shifted to just above the northern horizon. Since insects respond primarily to zenith polarization, they cease traveling at midday in summer and substitute other activities.* (Drawing by H. Peter Loewer)

the sky is directly overhead, in the zenith. In contrast, at solar noon in the Northern Hemisphere, when the sun is in the south and at its highest daily altitude, the antisolar point is far below the northern horizon, and the area of maximal polarization has shifted northward out of the zenith.

Consequently, zenith polarization waxes and wanes daily, with its maximums near sunrise and sunset and its minimums at noon. Seasonal differences in solar altitude affect the amount of that minimum. In summer, the unpolarized, glaring aureole surrounding the high midday sun intrudes into the zenith, even in the higher latitudes of the Temperate Zone, thus reducing overhead polarization to zero. Varying amounts of polarized light occur in the zenith at noon during other seasons, at least when the sky is clear. A sufficiently dense, unbroken patch of cloud entering the zenith at any time of day immediately eliminates polarization there for the duration of its passage.

Thus, there are two events leading to minimal values of zenith polarization: one recurs daily near solar noon during summer; the other occurs irregularly, but still predictably, with the passage of sufficiently dense clouds during any season. Together they provide a route for discovering how perception of zenith polarization affects the lives of insects.

Although several hypotheses have been proposed, we do not know how insects "see" polarized light. We do know that whenever polarization is sufficiently intense for them to do so, they adopt and hold a heading by referring to the plane of polarization while they travel. Their reliance on the plane can be demonstrated by holding a polarizing sheet, such as Polaroid, over them as they travel on the ground. If the axis of the filter is aligned so that it transmits light in the plane the insects are using, they ignore its presence and maintain their original heading. But if the sheet is suddenly rotated 90 degrees to change the plane of the transmitted light, the insects will abruptly change their heading by the same amount, slavishly following the shift in plane. In contrast to that typical response in the presence of polarized light, insects traveling during times of little or no polarization neither follow straight courses nor respond to polarizers rotated over them. They are not using the sky as a guide.

The presence or absence of a response by grounded insects traveling under Polaroid can thus be used to detect changes in their capacity for long-distance, straight-line travel. We can compare their responses at solar noon or when the zenith is cloudy with their responses at other times of the day. Simultaneous observations of the behavior of similar insects flying free in the vicinity will soon show whether there are any changing patterns in their actions that can be associated with the movements of the grounded insects under Polaroid.

The observations can take several forms. We can compare pairs or groups of the same or different species. In some tractable species, the responses of one individual can be compared during an alternating sequence of ground and aerial tests whenever the state of the sky changes. Meanwhile, if during the series of comparative tests we record all the various kinds of naturally occurring activity and behavior among insects in the vicinity, a picture of the changing patterns of behavior at the site soon begins to emerge. And if some insects are marked and released so they can be identified during later encounters, the evidence they provide about the proportions of residents and transients in the locality completes the picture. Such observations can be made wherever industrial smog has not totally obscured the sky; most of those described here were made in my garden.

Some especially interesting differences between residents and transients appear when the summer sun eliminates polarization from the zenith. Dragonflies, which are noted for their effortless flight and hawking skill, provide a dramatic example. When there is no polarized light in the zenith, those skills are apparent only in places where the insects have established "beats" between familiar landmarks. Caught aloft by the noon sun's aureole while they are dispersing through unfamiliar territory, these masters of the air bumble earthward as ineptly as any novice pilot whose instruments have failed. If they land where there is no food, they perch quietly during the entire time that the aureole is near the zenith. But if there is a swarm of gnats poised over a nearby bush, they will intermittently dart into it, catch one of the dancers, then return to the perch to eat it. Thus, when there is no polarized light overhead, they not only stop traveling through unfamiliar territory but also change their hunting style from ceaseless roving to sedentary ambushing.

Dragonflies are not the only predators to change in this way. The black-and-white bald-faced hornet is normally a restless forager, ranging widely in quest of prey. But caught far from its nest by the midday change in the sky or by sudden cloudiness, it too perches like the dragonfly.

A rhododendron bush in my garden—where numerous small flies, as well as hornets, rest through the summer noon—provides the setting for a deadly game, one whose rules change with the air temperature. If the noon air temperature is below 60 degrees Fahrenheit, all the insects bask quietly in the sun, seldom moving unless startled. Even then, they do not fly straight off across the open spaces of adjacent lawn (that type of flight requires a sky compass) instead, they merely flit to a nearby leaf.

On a cool day, the flies rest close to the middle of the rhododendron leaves, where radiant heat from the sunlit surface warms them by as much as 10 degrees.

above the air temperature. A fly resting in the middle of a long rhododendron leaf is not likely to be startled by shadows moving on the outer edge. Consequently, since the fly rarely moves, the hornets rarely notice it.

If the air temperature is between 75 and 80 degrees, however, the middle of a rhododendron leaf may become too hot for flies and hornets alike. Both then perch close to the tip, where the leaf temperature is only 1 to 3 degrees higher than the air temperature. In that position, the flies, more often startled by shadows, repeatedly flit from leaf to leaf. The hornets pounce from ambush on any flies that land near them.

Mosquitoes that frequent alpine regions by day also exhibit a comparable change in feeding behavior during midday or under zenith clouds. During such periods, they stop their long, roving flights, resting instead on vegetation and rising only when disturbed (or attracted) by something moving nearby.

Not all predators become sedentary when their sky compasses fail. Ladybird beetles and several kinds of small parasitic wasps habitually search places where there may be potential hosts, moving on if they find none. When there is polarized light overhead they alternately search and travel, but when there is none, they cease traveling and confine themselves to one place, repeatedly searching the plants until the sky changes and they can again travel.

These beetles and wasps are frequently used in the biological control of pests, and one wonders whether some releases have failed because the insects were liberated at the wrong time of day. If the weather is clear, releasing such insects under a high sun would ensure some local searching before the inevitable dispersal takes place. But releases early or late in the day, when there is more polarized light overhead, would increase the chances of rapid dispersal, perhaps beyond the area where the concentrations of pests would have helped the predators establish themselves.

An insect's residence in a locality no doubt confers advantages. When other guides are lacking, familiar landmarks enable insects to move from plant to plant through their home range. While wasps and bees have always been considered superior in this ability, during this study I saw that other kinds of insects also have a remarkably good memory for landmarks, as well as a well-developed "sense of place," which they most often demonstrate near midday.

Some insects, for example, make extensive rounds each morning and afternoon, but return to one special place on their route each midday, at about the time the aureole of unpolarized light is beginning to invade the zenith. Some muscid flies and some butterflies, such as the painted lady, merely bask after they arrive, occasionally flitting briefly among the plants. Others, like the cabbage butterfly, may not actually stop their rounds at midday, but instead modify their flight path to avoid crossing open spaces such as lawns, which they had crossed and recrossed all morning. Rather than continuing to fly across such openings at midday, they negotiate them by fluttering slowly around the edges, tracing right-angled paths where they earlier made a diagonal. A few species, such as the drone fly and the Lorquin's admiral butterfly, even return each day from their morning rounds to a particular station in the locality they favor. There they perch or hover, darting toward any landmarking bee that comes foraging at noon.

In these territorial species, members of the midsummer generation may, in fact, behave aggressively toward almost any moving object. It is a startling experience to realize that the gaily colored insect that has been tapping at one's hat, spectacles, and camera as it circles persistently is not sociable but hostile. How should one respond to an infuriated butterfly?

During a three-week period in July, one Lorquin's admiral was exceptionally punctual according to the solar clock, arriving in my garden each morning within 2 to 3 minutes of the time the aureole began to invade the zenith. It entered the garden from the west, perched on the rhododendron, and vigorously responded to the silhouettes of passing bush tits and robins briefly limned against the sky, as well as to any bees or cabbage butterflies that were active at noon. It defended the same 15 by 15 foot space during each midday period, leaving the garden again in midafternoon. Finally, old and tattered, it was displaced by a younger admiral that arrived at the same time one day and drove it from the garden.

Well before solar noon on summer days, the drone flies and their relatives the bulb flies establish small territories, often no larger than 2 by 3 feet and rarely bigger than 4 by 8 feet. Within these boundaries, the flies dart aggressively toward anything that moves. Every arriving or departing bee is harassed and butted. Bumblebees tend to ignore the onslaughts and are rarely knocked off balance unless they are struck while leaving a flower. Honeybees on occasion may be driven from the site, but some of the smaller bees, leafcutting bees, for example, may be effectively barred from pollen collecting in the locality.

Although drone flies occasionally feed from the flowers in their territory, the bulb flies almost never do. All the same, the latter's proprietary behavior, more often than the drone flies', seems to prevent other insects from pollinating the briefly flowering plants near their stations.

During the late spring and early summer, some of these flies may occasionally behave aggressively when there is polarized light overhead, as well as when there is none, but the most general, persistent, and violent displays take place when the aureole is in the zenith at midday or when there is an overhead cloud cover.

Perhaps the increased aggression in the absence of polarized light is linked with the behavioral change that occurs then: the flies remain constantly alert within their small territories, instead of occasionally flying away to other parts of the garden for lengthy sessions of preening or feeding.

The spring and summer generations of all kinds of daytime fliers are most regularly affected by the midday minimum in polarization. Wasps, when they are far from their nests, cannot home then without a sky compass. Foraging bees are equally affected. If they have not learned the landmarks on their homeward route, they are confined to the collecting site until overhead polarization returns. Bees homing by polarized light are easily distinguished from those using landmarks. When the insects use landmarks, they fly low, zigzagging homeward from one route marker to the next. When they steer by the sky, however, they spiral upward briefly after taking off, suddenly adopt the appropriate heading, then maintain it by flying over, instead of around, trees, houses, and other obstacles.

As the summer advances, the aureole of the declining sun intrudes less and less upon the zenith and ultimately does not come near it. Thereafter, overhead polarization, although reduced, is not eliminated at midday, and the insects behave differently from their early summer forebears. By late summer or early autumn, the middle of the lawn at midday is no longer the quiet place it was earlier in the summer. Rather, there is as much din from rapidly passing insects at midday as at other times, and open spaces are no longer the noon preserve of local bees that know the landmarks.

When there is more travel during the middle of the day, there is no overt territoriality among butterflies or flies. The late summer generation of the drone fly is not aggressive. The flies continue their rounds through midday, feeding peaceably alongside other insects.

Clouds interrupt zenith polarization whenever they drift by in any season, and travel stops whenever they do. During the autumn, however, cool air temperatures prompt insects interrupted in flight by zenith clouds to bask in any available sunlight, just as they did on cool days earlier in the year. So periods of zenith cloudiness in cool weather may be periods of comparative quiet, when no insects fly. In contrast, insects that stop traveling on a warm day often continue to flit and buzz about one small area, thus producing a localized sound out of all proportion to their size or number. On a very hot day, that constant buzzing—made perhaps by only one nearby insect—sometimes leads us to believe that there is a great flurry of activity. This may be so, but more often the increased sound merely indicates that an insect has been forced to stay where it is too hot to sit.

I have repeatedly emphasized overhead polarization, ignoring polarization from other parts of the sky and also ignoring the solar image that insects supposedly use in sun-compass reactions. In so doing, I have followed the example of the insects I have watched, for they ignored the sun when they flew across open spaces. In the morning and afternoon they would continue to make long, straight flights as long as the zenith was clear, whether or not the sun was visible. But as soon as a cloud covered the zenith, even though the disk of the sun remained completely visible and even when it was not very high, they merely sat or flitted. If ever they could have used a sun compass, those were the times when they should have done so. Since they did not, I believe that they could not.

Furthermore, although it seems reasonable to expect animals with such large compound eyes to be able to use polarized light from parts of the sky other than the zenith, neither crawling nor flying insects seem to do so. In several experiments I have tried to induce crawling insects to respond to polarized light from parts of the sky other than the zenith, but they reacted to rotating Polaroid aimed toward those other parts only when their crawling surface was tilted so that the light from the new direction was "overhead."

Insects in free flight also behave as though the only visible sky were directly overhead. So whether or not their multifaceted eyes record events at a distance from the zenith, the insects' behavior shows clearly that they do not use information concerning polarized light from the lower reaches of the sky. Their important navigational aid is directly overhead.

Watching insects when they cannot travel gives a glimpse into a world where survival may be determined by an individual's position on a leaf, and where territories that are abandoned each afternoon are reoccupied and fiercely defended again the following noon. With a little patience, it is not difficult to observe these occurrences. But while doing so, we must struggle to imagine the processes whereby an inhabitant of that "polarized" world varies its territorial urge or changes its predatory style in response to visual signals that we can scarcely detect, let alone decipher.

2

The Shark's Sixth Sense

ADRIANUS J. KALMIJN
and
KENNETH JON ROSE

Three hundred million years ago, long before dinosaurs walked the earth, sharks were present. They remain abundant today and have hardly changed. They are not just remnants of some bygone era but are well adapted to our modern oceans. In part, their adaptation may result from being endowed with an extraordinary sense that is among the most remarkable in the animal kingdom—sharks can find their prey and orient themselves in the open sea by the detection of electric fields.

More than forty years ago, Sven Dijkgraaf, a sensory biologist working at the University of Utrecht in the Netherlands, began to suspect that sharks might have electrical sensitivity. His first clue came in 1934 when he noticed that in experiments with small, blindfolded sharks, they reacted to a rusty steel wire. When the wire was brought within approximately one foot of a shark's head, the fish would suddenly turn away. To establish whether sharks would react in a similar manner to a nonmetallic object, Dijkgraaf repeated the procedure with a glass rod. The sharks ignored it until the rod actually touched their skin. Only then did they move rapidly away.

The sharks' reactions seemed to duplicate the experimental results Dijkgraaf had read about in a paper published in 1917 by two Harvard University scientists specializing in the sense organs of fish. Working with the common brown bullhead, a species of freshwater catfish, the Harvard researchers had tested the responses of blindfolded fish to metallic and nonmetallic objects. The catfish reacted then as Dijkgraaf's sharks did seventeen years later, responding to the metallic rod when it came within a range of one to two inches and to the nonmetallic rod only on contact. In subsequent experiments, the two scientists showed that the reactions of the catfish were due, not to movements of the hand-held metallic rods or to odor particles emanating from them, but to electrical currents given off by the metal objects upon contact with water. Because of the similarity of his results to the Harvard studies, Dijkgraaf assumed that his small sharks were responding to the electric fields produced in the water by the rusty wire.

In 1951, long after these early observations, a British sensory biologist named Lissmann studied those species of tropical fish of the Amazon and the Nile that have special organs that generate electric fields in their environments. He was less concerned with such animals as electric eels and electric rays, which produce strong electrical discharges in order to stun their prey, than with species that create much weaker fields that are too feeble to be harmful either to their natural prey or to other animals.

Lissmann found that many South American knifefish and weakly electric fish from the African continent emit pulsed discharges continuously throughout their lives. To explain the function of these fields, he suggested that these fish probe their environs electrically. That is, they monitor their own fields and are able to detect nearby objects through the distortions those objects cause in the electric fields the fish themselves generate. Lissmann proposed that weakly electric fish such as the knifefish evolved from what he called a preelectric fish, meaning a primitive one lacking electricity-generating organs but possessing receptors sensitive to electric fields. Today the concept of a hypothetical preelectric fish is quite believable, for we know that catfish and sharks are acutely sensitive to weak electric fields but lack the organs to create their own electrical milieu.

Just how sensitive sharks are to weak electric fields has been demonstrated experimentally by Kalmijn. He found that within the frequency range of direct current up to about 8 hertz, sharks respond to fields of voltage gradients as low as a hundred-millionth of a volt per centimeter. That would be equivalent to the field of a flashlight battery connected to electrodes spaced a thousand miles apart in the ocean. Little wonder that this sensitivity is the highest known in the animal kingdom.

Sharks owe their astounding sensitivity to electric fields to several hundred sensory organs in the head region called the ampullae of Lorenzini. These organs are marked by pores in the skin of the animal's snout, leading to jelly-filled ducts that terminate in sensory cells. Scientists have known about the ampullae of Lorenzini for years, but their function had long been a mystery. Early researchers believed the organs to be sensitive to pressure, perhaps informing the shark about water depth. But that hypothesis appeared to be fallacious and was soon rejected. Later, the ampullae were found to be temperature sensitive, but this did not seem to be their function either. Finally in 1963, Dijkgraaf and Kalmijn, who were then working together, established that sharks use the ampullae of Lorenzini to detect weak electric fields. This discovery solved the mystery.

Once the function of the ampullae had been correctly identified, a basic question still remained: How do sharks make use of their electroreceptive capability?

Virtually every living creature unintentionally produces in water an electrical field so weak it can be detected only by the most sensitive instruments. This results from differences in the electrical potential of the skin, which varies from one bodily area to another. Because water is an electric conductor, these differences produce electrical currents. For example, in the fish that sharks take as prey, the mucous membranes lining the mouth and the gill epithelia in the pharynx create direct-current fields that fluctuate with breathing movements. These currents flow through the water along field lines around the animal and, in accordance with the laws of physics, become rapidly weaker and weaker with increasing distance from the animal. Even so, the voltage gradients emanating from small fish and wounded crabs, the main diet of many sharks, produce within a distance of about one foot, bioelectric fields well within the shark's sensory capabilities.

These findings suggested that sharks may use their electrical sensitivity to locate their prey. In 1971, Kalmijn began laboratory experiments designed to test that hypothesis. He put a small, live flounder in a sandy-bottomed tank containing hungry sharks of the species *Scyliorhinus canicula*. Flounder and other small fish are part of the normal diet of these sharks. The flounder swam to the bottom of the tank and buried itself in the sand. Then Kalmijn added a few drops of odoriferous liquefied fish extract to the water to arouse the sharks. (Sharks will go without food for days until they are motivated and attracted by its smell.) Excited by the odor of the extract, the sharks began frantically searching for food. When they came close to the buried flounder, they made well-aimed dives at it, uncovered it from beneath the sand, and devoured it voraciously. Since the flounder was hidden and the fish extract was diffused throughout the tank, it would appear that the sharks must have used cues other than visual and olfactory ones to find their prey. Might the sharks have located the flounder by intercepting its electric field?

Sharks not only have a keen sense of smell but, like most aquatic animals, they are also sensitive to minute movements in the water. They perceive these movements with microscopically small organs in the skin along a visible band, called the lateral line, that runs the length of the body. It was possible, then, that the sharks might have sensed the hidden flounder either by its odor or its movements. The next step was to make sure that the sharks had not sensed the hidden flounder merely by these means.

An enclosure consisting of agar, a stiff, opaque, gelatinous substance, was put around the flounder. Agar is a pretty good barrier to odors and water movements but permits the transmission of electrical fields. Under these experimental conditions the sharks continued to make well-aimed attacks at the buried flounder from the same distance and in the same frenzied manner as if the prey were not shielded by the agar. To make sure that the thick agar roof of the enclosure was really effective in stopping any possible odor from leaking out, the live flounder was exchanged for cut pieces of fish, which produce an even stronger odor but do not generate electric fields. If odor were to come through, the sharks would find the bait within minutes. Instead they swam over the agar chamber without showing the slightest interest in the food. Obviously the agar was odorproof.

To establish beyond a doubt that the electric sense was the means by which the sharks had located the live flounder, the agar enclosure was covered with a plastic wrap. This served as an insulator that would contain any electrical field within the enclosure but would not impede any water movements and vibrations that might have passed through the agar. Under these conditions, not a single shark noticed the live flounder. Electricity clearly seemed to be the medium the sharks were using to find their prey.

As a final proof Kalmijn simulated the presence of the flounder in the tank by passing a weak electrical current between two electrodes buried in the sand. The effect was dramatic. Motivated by a few drops of fish extract spread in the tank, the sharks charged at the electrodes as if they were the real prey. They dug away the sand and bit them. Responding again and again when swimming over the electrodes in their search for food, the sharks left the site only after finding it really contained none. So tempting was the artificially generated electric field that, when a piece of odor-producing fish was placed a short distance from the electrodes, the sharks, although attracted by the odor, continued to dive at the electrodes rather than at the fish.

The aforementioned electrical tests were all conducted in the laboratory where the scientist can control the environment. However, captive sharks in a tank might react differently from sharks in their natural habitat, where many stimuli affect them. The authors, Kalmijn and Rose, therefore decided to verify the laboratory data by testing wild specimens roaming freely in the ocean, even though that would present many problems.

Human beings swimming in the sea produce their own electric fields, which sharks can detect from distances of up to about three feet. The fields produced by metals carried or worn on the body, scuba gear, for example, are usually even stronger. Thus, when studying sharks in their natural environment, we would have to get close enough to work with them without introducing electric fields from our bodies or from metallic equipment we might use in the water.

During the summer of 1976, while fishing off Cape Cod, Massachusetts, we learned that the smooth dogfish shark, *Mustelus canis,* regularly frequents the shallow, in-shore waters of Vineyard Sound on its nightly feeding excursions. This predatory shark is a warm-season visitor from equatorial waters, which arrives north in May and returns south in late October or shortly thereafter. It is an active bottom hunter, preying on small local fish as well as on crustaceans and other invertebrates. The females reach an average length of slightly over three feet, the males are somewhat smaller. We decided to use this shark for a series of nocturnal studies.

To observe the animals' feeding behavior, we had designed a nonmetallic field setup that would reproduce our laboratory tests in a natural setting. Using a triangular pulley system, we ran a length of rope and plastic tubing from a rubber raft to a bare patch of sand seven feet below us and back to the raft. In the center of this system was an opening in the plastic tube through which we could release fish extract as an odor source. One foot to each side of that opening were two sets of electrodes that would simulate a flounder's electrical field.

After dispersing several chunks of herring (a readily available substitute for flounder) over the area, we began to pump small amounts of liquefied fish through the tubing to attract the sharks. At the same time, we turned on one set of electrodes, leaving the other off as a control. The otherwise dark area was illuminated with an insulated underwater light and we made our observations through a glass-bottomed viewing box. The glow in the water extended for some twenty feet around the raft but, surprisingly enough, seemed neither to attract nor scare the sharks.

The first animal to appear on the scene was an American eel attracted by the odor of the herring extract. It wandered in and nibbled at the odor source, ignoring the electrodes. Soon a much larger animal swam into the cone of light. It was *Mustelus canis.* Within minutes, another shark appeared. Together they circled the illuminated area, apparently trying to locate the odor-releasing, presumed prey.

Gradually the sharks began to zero in on the odor source but before reaching it, they suddenly turned sharply toward the current-passing electrodes, biting and thrashing at that portion of the rope. The hole from which the fish extract was emanating and the control, or "dead," electrodes were ignored. During several nights of research, hundreds of sharks were observed, and all of them responded in this way.

These observations clearly demonstrate that sharks, roaming freely in their natural habitat, can detect and take their prey by the exclusive use of their electric sense. The experiments also show that within the range of one foot, electrical fields override the vague odor cues that initially arouse and attract sharks from a distance of many yards. The odor of a wounded prey lingers in the water long after the prey has gone. Therefore, when following an odor trail, sharks ultimately depend on a more localized and precise cue to spot their prey accurately and to seize it with one quick move.

In addition to using its electric sense to find its prey, it is possible that sharks rely on this sense for other purposes as well. As already noted, the smooth dogfish sharks we observed in the waters off Cape Cod migrate southward during the winter. They must therefore be endowed with a good sense of direction. Kalmijn believes that this sense may be electrical and may derive from the animals' ampullae of Lorenzini.

When a fish swims through the earth's magnetic field, an electrical field is induced. The polarity of these induction fields depends on the direction in which the animal is moving. Reception of these fields through the ampullae of Lorenzini may thus indicate to the animals the compass direction they are following. Sharks, in a manner of speaking, may have an internal electromagnetic compass.

The shark's ability to orient with respect to the earth's magnetic field has already been demonstrated in laboratory experiments. To verify these new findings and extend our earlier observations, we are currently outfitting a mobile, nonmetallic working platform (a modified Boston Whaler) for comparative studies of the electric sense not only of shallow-water, bottom-dwelling sharks but also of the open ocean, pelagic species of the Cape Cod area.

Although sharks have recently received a spate of bad publicity, most species are, in fact, harmless to humans. It is our contention that the shark should be respected, if not admired, as an ancient survivor possessing a special sense not acquired by most animals.

3

Night Fighters in a
Sonic Duel

KENNETH D. ROEDER

Moths are one of the main food sources of certain families of bats. They are attacked on the wing and in darkness in a contest in which speed and maneuverability are the premium qualities. That this nocturnal "game" has probably continued for some millions of years tells us that the contest is balanced: all bats locate and capture some moths; some moths detect and evade all bats.

While in flight, insectivorous bats emit a series of brief chirps pitched several octaves above the highest note audible to human ears. Each chirp is an ultrasonic tone that lasts only a few milliseconds. In many bats the tone drops in pitch by about one octave during this brief interval, so that if it were audible to us it would sound very much like the chirp of a bird. A bat makes these chirps about ten times per second cruising in the open; if it encounters any object in its flight path, its chirp rate may go higher than 100 per second.

At Harvard the precise and ingenious experiments of Donald Griffin and his students have shown that echoes returning to the ears of the bat inform it in detail about the size, distance, and location of objects in its flying path. The world of a flying bat must be a series of single and multiple echoes of a subtlety that we still do not completely appreciate. If a man walks about opening and closing his eyes rapidly, the visual world becomes a series of still pictures interspersed with intervals of darkness. However, in the bat's world discontinuities in perception are far more complicated because sound travels extremely slowly compared with light. For a bat, the spatial dimensions of the visual world are temporal dimensions in an acoustic world; a flying moth becomes an intermittent, fluctuating point in time.

About 100 years ago it was suspected that moths could evade bats through a sense of hearing. The sonar system used by bats was then unknown, so this was a truly inspired guess. Since then, studies of the anatomy of the tympanic organ in various species and families of the Lepidoptera, and observed changes in the behavior of moths in the presence of man-made ultrasound, have confirmed the suspicion that members of certain moth families can hear the chirps of echo-locating bats. Several families of moths possess tympanic organs, including the largest families of common, medium-sized moths—the Arctiidae, Phalaenidae, and Geometridae. It seems probable that tympanic organs have evolved more than once.

Dr. Asher E. Treat of the City College of New York (*Natural History,* August-September, 1958), first introduced me to the moth ear, and we worked together on tympanic nerve experiments. Field experiments with free-flying bats were carried out at his summer home in Tyringham, Massachusetts. Without his enthusiasm and skill in dissection we probably never would have tried to discover the defensive role of moth hearing.

The ear of a moth may seem to be a somewhat esoteric subject for a study of the form in which environmental information is coded in nerve impulses. But in some families of moths, notably the owlet moths, or Noctuidae, the tympanic organ contains only two acoustic sense cells. Electrodes placed on the tympanic nerve containing the axons (impulse conductors) from these sense cells can intercept all of the impulse-coded information this sense organ is capable of delivering to the moth's central nervous system.

The ear of noctuid moths is found on the thorax near the "waist," where thorax and abdomen join. A thin eardrum, or tympanic membrane, is directed obliquely backward and outward into a cleft formed by flaps of cuticle, and is normally covered by a thin layer of fine scales. Viewed from outside, the tympanic membrane often shows interference colors, indicating its extreme thinness.

Dissection under a microscope shows that the tympanic membrane forms the outer wall of the tympanic cavity, which is an air-filled, expanded portion of the moth's respiratory system. A fine tissue strand, the acoustic sensillum, is suspended across this cavity, and is supported near its midpoint by a minute ligament attached to another part of the skeleton. The sensillum contains the pair of acoustic receptors, or sense cells. Each acoustic sense cell (*A* cell) bears a bine distal process ending in the scolops, a minute refractile structure that extends toward the tympanic membrane. From the central end of each *A* cell an axon (impulse conductor) passes within the sensillum toward the skeletal support; this pair of *A* axons continues in the tympanic nerve to the thoracic ganglia. Passing the skeletal support, the *A* axons lie close to a large, pearshaped cell (B cell) that may have numerous fine, finger-like extensions reaching into the surrounding membranes. The B cell gives rise to a larger axon that runs parallel to the A axons in the tympanic nerve, eventually reaching the central nervous system.

An experiment will usually begin with the capture of a noctuid moth— perhaps one of the common army worms whose larvae do so much crop damage or, better still, a larger red underwing. Under temporary anesthesia, the moth is decapitated and firmly restrained with small strips of Plasticine on the stage of a dissecting microscope. It is kept in such a position that the tympanic openings have an unrestricted sound field. The scales on the thorax are removed with a small paintbrush, and the dorsal part of the thorax, including one of the main sets (horizontal) of flight muscles, is dissected away. The tympanic nerves run forward on either side of the cavity thus revealed, passing from the tympanic organs at the back of the thorax to the large pterothoracic ganglion that supplies all organs of the thorax.

There are several nerves in this region, all small and transparent. However, the task of hooking a tympanic nerve on an electrode is not as hard as it might seem. One electrode is a silver wire inserted anywhere in the tissues of the moth. The other is a silver wire tapered to a fine point that is bent into a minute hook. This active electrode is manipulated mechanically. Both electrodes are connected to an amplifier and cathode-ray oscilloscope, and also to a loudspeaker. Since nerve impulses cause minute brief current pulses at the electrode, they can, when amplified, be made audible as clicks in the speaker. When the tympanic nerve has been hooked, the speaker replies to ultrasonic sounds with a rapid sequence of clicks. These same nerve impulses can be photographed when displayed as spike potentials on the screen of the cathode-ray oscilloscope.

When the silence is broken by continuous, pure ultrasonic tones of various intensities, the A cells respond immediately. At the onset of a very faint tone (1), one A cell generates a small burst of spikes that immediately tails off into an irregular sequence. At a higher intensity (2), the initial frequency of A spikes is greater, and a regular discharge continues during the tone, though with declining frequency. At a still greater sound intensity (3), the A-spike frequency increases again, but it still declines as the tone continues and occasional spikes appear to have double peaks. At the highest sound intensity used in this experiment (4), the nerve response becomes quite complex—there are many spikes, double peaks, and spikes that appear to have double the normal height. These extra spikes are generated by the less sensitive A cell. In all the records the much larger spike potential of the B cell appears infrequently but at regular intervals, and is completely unaffected by the ultrasonic stimulation.

This experiment demonstrates, for one, that the intensity, or loudness, of the tone is encoded in the tympanic nerve discharge as spike frequency in the A axons. The evidence also shows that faint sounds are detected only by one A cell, while louder sounds are detected by both. As the sound continues, a decrease in spike frequency takes place. This decrease with the passage of time must mean that the sound is represented to the moth as becoming progressively fainter, even though it has remained physically unchanged. Such a progressive loss in sensitivity is known as sensory adaptation and is actually widespread and familiar in everyday experience. If adaptation did not occur in most receptors registering changes in the outer world, the impact of our surroundings often would be unbearable. The brilliance of a lighted room entered after dark would remain blinding, and the contact of our clothing would irritate our skin the day through. The speed with which receptors adapt varies greatly: the moth's A cells adapt relatively rapidly; other sense cells adapt slowly, if at all.

In another experiment an ultrasonic pulse was generated artificially at regular intervals. It was similar to a bat chirp except that it lacked the frequency modulation of the natural sound. A microphone was placed near the tympanic organ of the moth. The intensity of the sound pulse was adjusted until it just failed to produce a response in the A cells. The intensity was then increased in measured steps of 5 decibels; the microphone and nerve response were recorded at each step. Part of the findings were already familiar, but there were two additional ways in which the A response changed as a result of increased intensity. Although each sound lasted only 3 milliseconds, at the higher intensities the spike discharge continued for several milliseconds after the sound had ceased. It was as if the more intense sounds caused in the sense cells some overaccumulation that then continued to generate impulses after the sound itself had stopped. Second, the response time (the interval between the stimulus and the first A spike) became shorter as sound intensity increased.

It is important to note that the above-mentioned A-cell properties are similar to those reported many times previously in many animals by many observers. Heretofore, however, such observations have dealt mainly with single units that were isolated for experimentation from a complex sense organ containing many thousands of receptors. The behavior of the A cells is significant because the cell represents the whole sense organ, not merely a small part of it. Therefore, the A cell defines the total sensory input being communicated to the effector mechanisms for the evasion of bats.

Other experiments with artificial sounds showed that the tympanic organ can detect sounds ranging from 2 kc/s (kilocycles per second) to as high as 150 kc/s. The upper limit of human hearing is 15 to 20 kc/s. Even with this great range, moths appear to be tone deaf. They seem to have no mechanism for discriminating one frequency or pitch from another; the tympanic organ is mainly concerned with discriminating differences in sound intensity, or loudness. With only one ear, a moth could measure loudness from the A-spike frequency, and from activity in one versus both A axons. By using both ears, the moth could "compare" two different hearings of the same sound, which might register with more intensity on one side than on the other.

These experiments with artificial sounds have introduced the elements of vocabulary and grammar of the neural language. Fortunately, the tympanic organ communicates with the moth's central nervous system only in the simplest form of this language, so that even after this elementary instruction it is impossible to interpret some biologically significant messages. These messages are the chirps made by bats in their natural occupations.

Our first record of a tympanic nerve response to the chirps of a flying bat was obtained in the laboratory, and almost by accident. Experiments with artificial sounds were in progress during January, a time of year when New England bats are deep in hibernation. A student making a weekend exploration of a New Hampshire cave found a hibernating bat and brought it back to the laboratory, where it was placed in a refrigerator and almost forgotten for several weeks. When removed and held in the hand of an experimenter near a tympanic-nerve preparation and microphone, the bat recovered sufficiently to deliver a few angry and audible shrieks, and an energetic bite. This last naturally brought about its release, whereupon it flew "silently" around the laboratory close to the ceiling. Throughout the flight the prepared tympanic nerve delivered a rapid series of short bursts of A spikes. When the bat flew sufficiently close to the experimental table, the microphone joined in with its electronic version of the ultrasonic chirps.

This impromptu experiment showed not only that the tympanic organ responds as expected but also that it is highly sensitive to bat cries. One or both of the A fibers continued to respond at times when the bat was too distant for its cries to register in the microphone. The moth could hear the bat at all points within the laboratory, and we were most eager to go beyond its walls and into the field.

This turned out to be somewhat more than the carefree jaunt it suggests. The next summer a load of about 300 pounds of electronic gear was hauled up a grassy hillside in the Berkshires of Massachusetts, and reassembled in a spot where bats were known to feed. At dusk a moth was captured at a nearby light and mounted so that one tympanic organ had an unrestricted sound field. Its tympanic nerve was hooked on an electrode and the A and B fiber activity was followed continuously on an oscilloscope and loud-speaker. Spikes were also recorded on magnetic tape.

The high excitement of listening for the first time to night sounds through a moth's ear was tempered by the thought that we had no independent evidence that they were being caused by bats. They were inaudible to us, and in this first field experiment we had with us no ultrasonic microphone to provide a separate record. A floodlight was rigged so that we were able to observe bats flying within 20 feet of the preparation. It then became clear that the range of the moth ear was much greater than that of the light, so that the appearance of a bat in the lighted area could often be predicted by listening to the rising pitch of successive *A* bursts from the moth ear.

It was difficult to establish the range of this biological bat detector, since it depended upon the species of moth and bat as well as the relative angle of their flight paths. In another experiment a moth preparation was set up at dusk about 200 yards distant from an old barn where bats roosted. It was known that at this hour the bats usually left the roost singly and flew on a straight path directly over the site chosen for the preparation to other feeding grounds. An observer, wearing headphones connected by a long cord with the amplifier, walked "upstream" toward the barn while listening for the first signs of regular *A* bursts from the moth ear behind him and watching the bats pass overhead. The maximum distance for *A* responses lay between 100 and 120 feet from the moth ear while the bats were flying toward the ear at an altitude of about 20 feet.

All information heard by the observer came from one ear of a moth. What could be learned by recording from both ears simultaneously? This project had to wait until the following summer, for it was necessary to learn how to insert and manipulate two hooked electrodes within the small space of a moth's thorax, and to duplicate most of the amplifying and recording equipment. The activity in right and left tympanic nerves was recorded on stereo magnetic tape and was subsequently photographed by replaying the tape into a dual-beam oscilloscope.

The bat's approach is initially signaled by a group of spikes (the first is a *B* spike, the rest *A* spikes). The second ear does not detect the bat until its next chirp, when the number of spikes indicates less intensity compared with spikes displayed by the first ear. This difference persists in the third response, but by the fourth there is little difference between traces. This suggests a bat approaching from one side, then moving directly overhead.

It is interesting to listen through stereo headphones to the taped responses of right and left tympanic nerves to a moving bat. The human ear interprets these spike differentials as giving direction to the source, and one can almost imagine oneself inside the nervous system of the moth as the source of clicks appears to move from one side to the other. This illusion of direction is not continuous, and much of the time the source of sound seems to be in the center of one's head. The explanation is that the spike differential is greatest at low chirp intensities, becoming less and disappearing above a certain loudness. This saturation of the acoustic response above certain sound intensities indicates that a moth would be better able to determine the bearing when a bat was near the moth's maximum range of hearing.

A differential response would be possible only if the ears of a moth were somewhat directional, responding better to sounds on one side of the body than on the other. A polar graph showed that, while there was little difference fore and aft, a click on the near side of the moth was heard at about twice the distance of a symmetrically placed click on the far side.

This information extracted from the tympanic-nerve responses makes possible a crude prediction of the moth's behavior upon detecting an echo-locating bat. If it is assumed that a bat is first detected at a distance of 100 feet and then approaches on a straight path at right angles to the moth's course while making chirps of constant intensity, the differential tympanic response would decline from a maximum at about 100 feet to zero at 15 to 20 feet. Within this range the moth would have sufficient information to enable it to turn away from the direct path of the oncoming bat. At a range of less than 15 to 20 feet the neural information reaching the moth's central nervous system would make possible only nondirectional responses vis-à-vis the bat's position.

This is as far as we can go at present in assessing the acoustic information coded and transmitted to the moth's central nervous system by the A cells. Until we know more about the anatomy and neurophysiology of the moth's pterothoracic ganglion and brain we must redirect our curiosity.

It is easy to show that some moths respond to high-pitched sounds, such as the squeak of a glass stopper, the jingle of keys or coins, the high notes of a violin or flute, and a variety of rustling and hissing sounds. But it is somewhat harder to describe just what they do. Some fold their wings and fall to the ground; the flight of others becomes faster and more erratic; some fluttering individuals become motionless; inactive moths may take flight.

Similar reactions can readily be observed in moths being chased by bats. As a bat comes "silently" out of the darkness the flight pattern of the moth suddenly changes to any one of a number of maneuvers—dives, rolls, repeated tight turns, or rapid flight just above the ground. The bat may make a single pass, or turn at once to make another, or it may attempt to follow the moth through its gyrations.

It is a dizzy "dogfight." Extrapolation of a string of acoustic dots in time is pitted against unpredictability; power and speed against maneuverability. The details may be difficult to discern, but the outcome is seen either as a bat and a moth going their separate ways, or as a departing bat and moth wings fluttering slowly to the ground.

We made an attempt to find out the extent to which the odds in this contest are influenced by the avoidance tactics of the moth. We observed 402 encounters between bats and moths and scored for the presence or absence of a sudden change in the flight pattern of the moth as the bat approached, and for the outcome—capture or escape of the moth. Analysis of the pooled data showed that for every 100 reacting moths that survived an attack, only 60 nonreacting moths survived. Thus selective advantage of evasive action was considerable.

Such procedures focus upon only one instant in the life of a moth, although certainly it is an important one. It is possible that at other times the possession of tympanic organs and evasive mechanisms weigh differently, even negatively, in survival, so this measure does not describe the overall survival advantage of possessing tympanic organs. Nevertheless, it could account for their evolution.

In passing, it is interesting to note that some species of moths are prone to infestation of the tympanic cavity by mites. These parasites have been found to infest only one ear, however, and by this behavior pattern appear to ensure their own survival.

It is not easy to tell at what instant a cruising bat first detects a medium-sized moth and turns to the attack. It seems unlikely that the bat makes acoustic contact at distances greater than 10 to 15 feet. The tympanic-nerve studies showed that within this range the average bat cry is capable of saturating both ears of a moth, so that the latter can make only nondirectional responses in attempting escape.

Yet the tympanic organs can detect a bat cry at distances of 100 feet and perhaps even more. At this range there is a marked difference in the nerve responses of the right and left ears when the bearing of the bat is to the right or left of the moth. There would seem to be little survival advantage to the moth in making erratic turns and twists when the predator was still so distant, although they could be of value at close quarters when the small inertial moment and short turning radius of the moth is pitted against those of the more massive bat.

The complexity of the natural situation, in which both sound source and detector are continually on the move, was reduced by replacing the bat with a stationary multidirectional transmitter of ultrasonic pulses. The transmitter was mounted on a 16-foot mast at the edge of a lawn surrounded by low vegetation. The observer was seated 25 feet behind a floodlight that illuminated a broad area of garden and silhouetted the transmitter on its mast against the night sky. This view of the transmitter on its mast against the night sky. This view of the transmitter was also framed in the field of a 35 mm. still camera.

The observer had at hand two switches, one controlling the ultrasonic signal and the other the camera shutter. When a moth was seen to move into the field of the camera the shutter was opened and the moth's track was recorded as a continuous line against the black background of the sky. Undulations on the line were caused by the moth's wing movements. After a stretch of flight track had been recorded the switch controlling the ultrasonic signal was depressed. This released a train of ultrasonic pulses, commonly at a rate of 30 per second, each 5-millisecond pulse having a frequency of 70 kc/s. The pulses were "shaped" as much as possible to resemble bat cries, but they lacked the frequency modulation of the bat's natural sound.

By the above means, the moth's flight path was recorded before and during ultrasonic stimulation. The onset of the pulse sequence is shown by an extra-bright spot on each photographic record, while the timing of events is indicated by gaps repeated at quarter-second intervals throughout the track.

The worst defect of the method is the large amount of light needed to secure a satisfactory photographic record. Light, we were afraid, might have altered the responsiveness of moths to a signal they normally encounter only in darkness. However, visual observers working with illuminations too low for photography, and with yellow and red light to which moths are much less sensitive than is man, reported no substantial differences in moth behavior. Another problem lay in the difficulty of identifying the moth species producing the tracks. Many flew away before they could be captured, while others dived into the vegetation and could not be found.

Moths that reacted within 10 feet or so of the transmitter showed a bewildering variety of reactions, usually ending in a dive, irrespective of whether the moth was above, below, or to one side of the transmitter at the time of the stimulus. The simplest reaction seemed to be an abrupt dive with wings closed. Moths reacting at a greater distance from the transmitter showed a distinct tendency to turn away from the source of ultrasound and continue in level, although often accelerated, flight.

Thus, the prediction of the neurophysiological observations seems to be confirmed by behavioral observations; high sound intensities produce nondirectional responses; low sound intensities result in directional responses. The great sensitivity of the tympanic organs must provide moths with an "early-warning" signal that prompts them to move out of the general area in which bats are feeding. As the number of impulses in the tympanic-nerve transmission increases to the saturation point the message may be thought to change to the "take-cover" signal, at which point the moths dive for the ground.

Like most biological observations, this one raises a dozen questions for every one it answers. Most of the moths making these tracks certainly belonged to the families Arctiidae, Phalaenidae, and Geometridae. How do the several families lacking tympanic organs, some containing common and successful species, survive without ability to hear bat cries? Recently, two British workers, D. Blest and D. Pye, have shown that certain tropical arctiids produce trains of ultrasonic clicks when teased or shaken. It will be interesting to see how this ability to make noises audible to bats fits into the contest between prey and predator. Tympanic-nerve responses recorded from different moth species are generally consistent and similar, except perhaps for sensitivity. On the other hand, the variety of nondirectional maneuvers released by high-intensity ultrasonic stimulation defies any attempt at orderly description. Does each species have its characteristic pattern of response? Or does it have a repertoire upon which it can draw in random order? Does sound intensity or some other sensory condition play a part in the pattern of response? There is some comfort in the thought that this unpredictability, however determined, is probably as confusing to the bats as it is to the experimenter, and therefore may also be of considerable importance in the survival of moths' evasive behavior.

4

The Pit and the Antlion

HOWARD TOPOFF

An insect larva is a transient container inside of which are developing all the important evolutionary adaptations of the adult organism. But a larva can lead an existence every bit as fascinating and complex as the adult. Even without such structures as wings and copulatory organs, an insect larva is a mature animal in many respects, often possessing specialized adaptations that endow it with a lifestyle markedly different from its adult parents. One of the best examples of such a developmental dichotomy is the antlion, which together with the dobsonflies, owlflies, and lacewings belongs to the order Neuroptera.

The adult antlion, with its long, slender body and delicate outstretched wings, resembles a damselfly. The larva, by contrast, is a roughly wedge-shaped creature that spends almost all of its time hidden beneath the soil. Because their specialized structural and behavioral adaptations make them efficient and dramatic predators, antlion larvae are among the few nonpest insects to have generated more behavioral studies than the adult forms.

What has made larval antlions a favorite study subject for more than 200 years is their behavior; they construct a funnel-shaped pit in dry sand and prey upon small terrestrial animals that accidentally fall into it. As its common name implies, the antlion larva feeds predominantly upon the many species of ants that forage in its environment, not because of any unique feeding specialization, but simply because ants are usually the most abundant wingless arthropods wherever antlions are found. The larva will, in fact, just as readily consume a wide variety of arthropods, including spiders, sowbugs, caterpillars, and beetles.

An antlion excavates its pit by moving backward in a circle, using its oval-shaped abdomen as a plow and its flat head as a shovel for flicking sand upward. When the pit is completed, the larva lies motionless on the bottom, concealed beneath the sand, with only its long, piercing mandibles exposed. When a foraging ant or other suitably small prey accidentally steps over the rim of a pit, the sand particles making up that portion of the sloping wall roll downward, carrying the victim to the skewerlike mandibles of the waiting antlion (figure 4.1).

Frequently, the antlion does not successfully impale its prey on the first attempt, so the prey tries to escape up the angled wall and out of the pit. But as the potential victim scurries upward, it dislodges the tiny dry sand particles beneath it and finds itself on a treadmill—moving its legs as fast as it can but making no upward progress. To make matters worse for the prey, the antlion responds to the mechanical stimulation of the miniature sand avalanche resulting from this activity by flicking its head upward and showering the prey with sand. As this storm of loose sand falls on the slope of the pit it speeds up the treadmill effect. The prey loses ground and eventually tumbles back toward the waiting antlion.

As soon as it has trapped the prey with its mandibles, the antlion moves backward, dragging the victim deeper into the sand and sucking out its protein-rich body fluids. Holding on to its prey is no simple task, especially when an antlion snares an animal considerably larger than itself. Here it is aided by a morphological adaptation consisting of numerous hair tufts that project anteriorly. Anyone who has ever used a toggle bolt will appreciate how these hairs function. When the antlion moves backward, the hairs bend and offer little resistance. But when a struggling prey organism attempts to pull the antlion forward, the hairs flare outward and anchor into the substrate.

For several springs, I have led an animal behavior class at The American Museum of Natural History's Southwestern Research Station in Arizona. We worked exclusively with *Myrmeleon immaculatus,* a species extremely abundant in the fine silt deposited along the banks of a creek near the research station. Many of our observations focused on the actual process of pit construction. We noted that before a pit is excavated, the larva moves just beneath the substrate, flicking sand continuously as it plows backward across the terrain. When a larva moves backward in this fashion, it leaves a narrow furrow in front of it, much as would be created by dragging the tip of your finger lightly across the sand. Because the path of the larva can seem tortuous, these movements are called doodles, hence another popular name for the antlion: doodlebug.

The success of both doodling and pit construction requires the antlion to overcome several formidable problems. Digging through sand with efficiency is one. The antlion accomplishes this by utilizing a variety of morphological specializations. Foremost among these is the shape of the larva's abdomen, with its relatively blunt anterior end gradually tapering toward the posterior—the direction

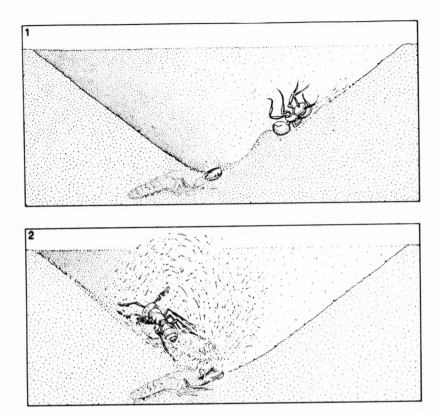

Figure 4.1. A wandering ant has stumbled into an antlion's pit. *Alerted by cascading grains of silt loosed by the ant, the antlion awaits its prey with poised mandibles (1). Having eluded its predator's first attack, the ant scrambles up the 38-to-42 degree slope of the pit in an attempt to escape. The antlion furiously flicks silt upward, preventing*

of movement. This shape enables the antlion to slide easily backward through the dense substrate when the hind legs move forward on their power stroke. The animal is also aided by its hair tufts, which bend forward when the animal moves backward, but which fan outward and anchor it when the legs move back on their return stroke. (This prevents the animal from simply oscillating back and forth in the sand.) In addition to functioning as anchors, one particular group of tufts on the ventral surface of the abdomen seems to have an additional function. These tufts occur along two longitudinal ridges and the hairs growing from them are longer than those on most other parts of the body. This arrangement actually creates two furrows relatively devoid of sand. As a result, when the larva's hind

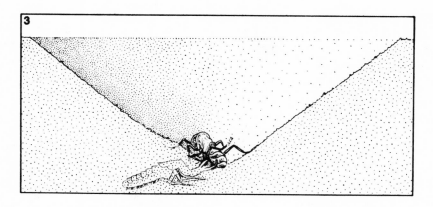

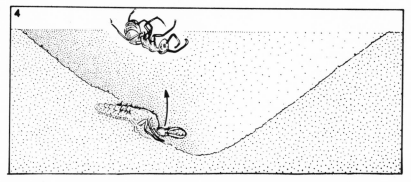

the ant from gaining purchase (2). After the ant is finally trapped, the antlion sucks out its body fluids (3). Finished with its meal, the antlion flings the prey's carcass out of the pit (4). (Drawing by Alan Iselin)

legs move backward on the return stroke, they move within these air pockets, thus encountering far less resistance than when moving through the sand on the power stroke.

Because pit construction is almost invariably preceded by some amount of doodling, biologists have always assumed that doodling is important in enabling the antlion to sample the environment and locate a suitable site for its pit. It has never been made clear, however, exactly which parameters of the environment provided the selection pressure for the evolution of doodling. It is doubtful, for example, that substrate type is the key parameter, because antlions are not terribly fussy about the medium in which they dig. The published literature shows that pits have been found in quartz sand, ted sandstone, dust, humus, rotted wood, gypsum, and even coal ashes. The only consistent requirement seems to be that the substrate be composed of small, dry, loose particles. Furthermore, because these variables are properties of the substrate itself, the larva is undoubtedly capable of responding to them without ever raising its head above the sand. But this is precisely the outstanding feature of doodling. As the antlion pushes its way backward through the sand, it repeatedly flicks sand upward, thus raising its head above the soil surface. It would make sense, therefore, if the primary adaptive function of doodling enabled the antlion to respond to environmental cues whose energy source was capable of stimulating receptors located on its head. The obvious candidates for this job are the well-developed eyes, which are situated on both sides of the head, close to the base of the mandibles. But what does the perception of light have to do with locating a site for a pit?

One indirect clue, pointed out in practically every field study of antlions, is that pits are seldom found in greatly exposed areas. They tend to be concentrated instead beneath trees or other natural overhangs. The advantage of such locations is greater protection from the sun, large animal traffic, wind, and rain. Although we cannot be certain at this early stage of the project, our studies have provided some circumstantial evidence that the larvae actively locate these sheltered sites by means of a visual orientation process that takes place during doodling.

When we released captured antlions within our experimental plot, much of the sand surface was covered with patches of shade produced by the surrounding trees and the early-morning sun. The larvae doodled in many directions during early morning, but all activity ceased throughout the hottest part of the day—late morning and early afternoon. Late in the afternoon, however, a second burst of doodling occurred. As the sun headed westward, the shaded areas on the sand surface gradually receded eastward. By this time, all the antlions were also doodling in an easterly direction, with some minor scattering, so that by sunset the experimental plot was covered with a series of relatively straight and parallel doodles. Some of the antlions were actually keeping pace with the moving shadows. Although it was not clear whether the antlions were responding directly to the darkened areas of shaded sand, to some aspect of the leaf and branch pattern from the trees, or to temperature gradations, they were all clearly moving toward the shaded locations.

Sometime between 5:00 and 7:00 P.M., each antlion abruptly ceased doodling in a straight line and began to move in an almost perfect circle. Pit construction had begun. As each antlion circled backward, it continued to flick sand up over its body and out of the circular furrow. When it completed the first circle, it moved inward and continued along another circle of slightly smaller diameter. The larvae continued this pattern of movement until an inverted cone was formed, then it took up a predatory position and ceased activity.

An antlion often finds its backward movement during pit excavation blocked by small stones, pieces of twigs, and other debris. One response to this situation is for the antlion to alter its course slightly and simply bypass the obstacle. Unless the stone is exceedingly massive, however, there are several reasons why this is not the most effective adjustment. If the antlion shifts its circling pattern outwardly, around the pebble, it will only encounter the pebble again while excavating the interior of the pit. On the other hand, if the antlion bypasses the stone by moving inward, the obstacle winds up perched near the rim of the crater. Unfortunately, this turns out to be only a temporary solution because the action of wind and passing animals could easily cause the pebble to fall back into the pit. In addition, obstacles around the rim of the pit may cause potential prey organisms to alter their direction away from the pit.

In order to study the antlion's response to obstacles, we collected several dozen larvae, ranging in weight from 14 to 54 mg, and placed them in coffee cans filled with fine sand from their field environment. Each evening when the animals were engaged in pit construction, we carefully placed pebbles of varying weight directly in their paths. We found that the strategy of altering the direction of circling is indeed used by the antlion, but only when the obstacle is so massive that it cannot be budged. When this occurs, however, the antlion doesn't merely move around the pebble. Instead, it actually stops circling, doodles to another area several centimeters away, and excavates an entirely new pit. If the obstacle is at all movable, however, the antlion promptly removes it. We discovered that the *Myrmeleon* larva possesses at least four different mechanisms for removing pebbles from its pit, depending upon how heavy the pebble is in relation to its own weight.

Very small pebbles, weighing from a fraction to approximately five times the weight of the antlion, are treated like ordinary grains of sand. The antlion moves backward under the pebble until its head is positioned directly beneath and then flicks the obstacle out of the pit by catapulting it upward and back over the long axis of its body. Because the path traveled by the pebble is along a tangent to the circular groove in which the antlion is digging, we have termed this mechanism of pebble removal a "tangential flick" (figure 4.2A).

A slightly heavier object, between five and eight times more massive than the antlion, elicits a noticeably different response. Although the antlion assumes the same position as for a tangential flick, it now heaves the pebble laterally, perpendicular to its body axis, away from the center of the pit. The adaptive value of this "lateral flick" is that the pebble now has to travel a shorter distance in order to clear the outer wall of the pit (figure 4.2B).

If the pebble is approximately ten times the weight of the larva, the insect switches to a third mechanism, one that we call a "radial flick" (figure 4.2C). The

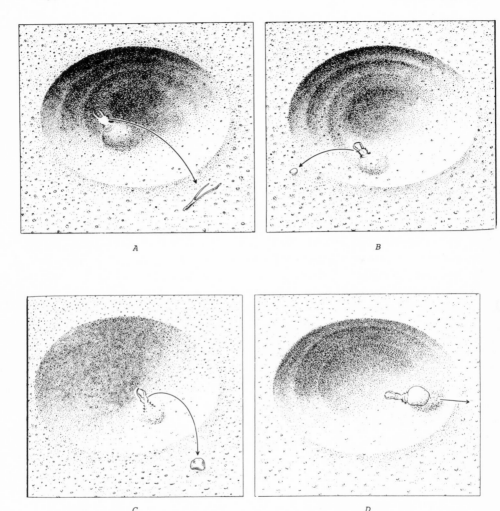

Figure 4.2. Pit obstacle-removal techniques.
A. Tangential flick for obstacles up to five times the antlion's body weight.
B. Lateral flick, five to eight times body weight.
C. Radial flick, ten times body weight.
D. Radial push, more than ten times body weight.

name of this procedure stems from the fact that the antlion now rotates its body 90 degrees, lining itself up along the radius of the circle, with its head facing toward the center. In this position, as in the previous lateral flick, the pebble is thrown out along the shortest path possible. The advantage to the antlion of taking this position for heavier pebbles is that by heaving the stone backward along its body axis, the animal is using its most powerful stroke.

The radial flick is the last mechanism available to the antlion for catapulting objects out of the pit. When it encounters an obstacle that is heavier than ten times its own weight, it initially makes several attempts to toss it into the air. But if the animal's musculature cannot provide the necessary power, the pebble merely jiggles slightly on the surface of the sand. After several futile attempts at flicking, the antlion emerges from the sand, places the tip of its abdomen against the stone, and slowly pushes it radially up the slope of the crater and out of the pit (figure 4.2D), stopping only after the pebble is several centimeters from the rim. Then the antlion does an about-face, moves back down the slope, reenters the sand, and resumes excavating.

Observing an antlion remove a pebble by means of this radial push is reminiscent of the legend of Sisyphus, King of Corinth. In Greek mythology, Sisyphus was the craftiest of men. When Death came for him, he instructed his wife not to offer the traditional sacrifice to the dead. Upon arriving in the underworld, however, Sisyphus complained that his wife was not doing her duty. He even managed to convince Hades, god of the underworld, to permit him to return to the upperworld to punish his wife. It was all a trick, of course, and when Sisyphus returned he and his wife lived together for many more years. After entering the underworld for the second time when he died of old age, Sisyphus was handed a most unusual punishment. His task consisted of rolling a huge rock up and out of a deep crater. Each time the boulder neared the rim, however, it rolled back down to the center of the crater and he had to start over. Thus Sisyphus's task lasted an eternity.

An antlion is not usually confronted with precisely the same problem Sisyphus had. Because most naturally occurring stones are irregularly shaped and have somewhat flattened surfaces, they do not automatically roll down into the pit if the animal momentarily loses contact. The antlion's task is also simplified in that the radial push need not be executed in one continuous movement. Instead, the animal might push the stone partway up the slope, stop for several seconds, and then resume pushing.

We decided to put the antlion to the ultimate Sisyphean test by repeating the obstacle experiment, this time using perfectly spherical ball bearings. The result was an exhibition of balancing that would be the envy of any circus seal.

Once again, the antlion emerged on the surface of the sand after several unsuccessful attempts at flicking a heavy steel ball. It placed its abdomen against the ball and started pushing slowly up the slope of the crater. Almost immediately, however, the smooth ball began to shift position so that the tip of the larva's abdomen was no longer applying force through the center of the ball. The antlion responded with a series of compensatory movements, similar to those one would make while balancing a broomstick vertically on the tip of an outstretched finger. Whenever the ball slipped either toward the right or left, the antlion reacted by shifting its abdomen in the same direction just far enough so that it was properly recentered on the ball.

When we repeated this experiment several times, it became obvious that we had just about approached the limit of the antlion's capability for behavioral adjustment. As the animal pushed the sphere up the slope, its abdomen was in an almost constant state of movement. More times than not, however, the larva lost control, the ball rolled down to the center of the pit, and the larva, like Sisyphus, had to start again.

Although placing pebbles and ball bearings in the paths of antlions during pit construction may at first glance seem to be little more than clever tricks designed by scientists to amuse themselves, these experiments can answer questions of fundamental importance. Animal behaviorists often use the term "stereotyped" when referring to the behavior of an organism whose capacity for adjusting is relatively limited. A serious problem arises, however, when this conclusion is reached as a result of studies conducted only in the species' natural environment. On the one hand, it may indeed be true that the organism simply does not possess the neural complexity necessary for a wide range of behavioral flexibility. On the other hand, it is also possible that novel situations arise infrequently, so that the organism only occasionally has the opportunity to demonstrate its complete range of behavioral patterns. This is one of the principal rationales for conducting carefully controlled experiments in a laboratory setting. By testing the organism under a variety of experimental conditions, the investigator can often elucidate, in a short time, a range of behavioral processes that might take many months of field study to uncover. Although our class experiments represent only the very first stages of a thorough research program, we have already demonstrated that the antlion *Myrmeleon immaculatus* possesses a far greater potential for behavioral adjustments than had previously been reported.

Although the excavation of pits by antlions has often been compared with the building activities of other arthropods, such as spiders, caddisflies, and honeybees, our findings concerning the antlion's behavioral plasticity suggest that the analogy is a poor one. Consider, for example, the beautiful cartwheel pattern of the orb web built by *Araneus diadematus,* one of the master weavers of the spider world. If the spider merely moved from one support to another, haphazardly depositing threads of silk along the framework, the result would undoubtedly be little more than a silky mass (and probably a silky mess). The formation of an optimal snare requires *Araneus* to utilize a precise sequence of its own intrinsic behavioral patterns and not rely on being helped by factors in its external environment.

In the case of the antlion, by contrast, the evolutionary process has taken a very different direction. The success of the antlion's behavior is guaranteed precisely because the organism is adapted for taking maximum advantage of the physical properties of the substrate in which it is excavating. As a result, its behavior can be quite variable, yet still produce an inverted conical pit with ideal trapping properties.

To understand how this works, consider the following simple exercise, one that anyone can perform at a sandy beach. A cylindrical pail is filled with the dry sand found far from the water's edge and carefully inverted on a flat surface. The resultant mound of sand will not have the cylindrical geometry of the pail. Because the sand is dry, particles slide past each other as soon as the pail is lifted, so that the mound will be relatively cone shaped. The angle by which the wall of the cone slopes upward from the base will always be similar as long as the same substrate is used. In earth science, this angle is termed the "natural slope" and represents the steepest angle under which an unbraced wall of material will remain in stable equilibrium.

Although this example is for the formation of a cone whose apex rises above the substrate, the same geologic principles are operative when sand is removed during the formation of a funnel-shaped pit. To demonstrate this, we collected several buckets of silt from the antlion's habitat, and with a small vacuum cleaner sucked up particles until we created a pit with a diameter comparable to that of a larva's pit in the field. We then measured the slopes of the walls of our artificial pits and found that their range was between 39 and 43 degrees. When we measured the slopes of antlion pits in the field, the calculated range was from 38 to 42 degrees. Our conclusion is that many of the features of an antlion's pit result as much from the action of physical forces operating on the silt particles as from the behavior of the antlion itself. Thus, it is not surprising that the antlion exhibits no rigid sequence of movements during pit construction. The antlion may excavate the initial circle by moving either clockwise or counterclockwise, and the direction of circling may change several times before the pit is completed. Another common variation is for the antlion to stop circling altogether and cut straight back and forth across the center of the depression, only to resume circling after an unpredictable number of such crossings. Finally, the antlion may simply spend varying periods of time in the center of the depression and flick headfuls of sand without changing the position of its body. As long as the antlion remains within the area delineated by its initial circle and spends enough time tossing out sand, the result will always be a first-rate trap.

5

Bird Navigation

Travels Around New England in Pursuit of Pigeons

CHARLES WALCOTT

Eels, salmon, whales, turtles, and birds—and even insects—migrate long distances, sometimes for thousands of miles. While this phenomenon has been studied throughout recorded history, we have as yet no clear answer to the question, How do animals navigate?

The most familiar of these migrating animals are those that perform the most spectacular migrations: the birds. The arctic tern, for example, sometimes flies 11,000 miles each way from the Arctic to the Antarctic, and back, each year. Closer to home, we are all familiar with the northward migrations of robins and bluebirds in the spring. These birds, having spent the winter in the warmer climate of the south, migrate north with the coming of spring. Interestingly, banding birds has shown that many return to the exact place where they were banded: the robin on your lawn this year was probably there last year and will most likely return next. This suggests that migration is far from a haphazard movement north or south. What we would like to understand is how the robin finds its way south in the fall, then returns the following spring to exactly the same place that its journey began.

That most of these birds migrate at night does nothing to make the problem easier, but it does suggest that the birds might be using the stars. And so it seems to be. E. G. F. Kramer and E. M. Sauer in Germany and Steven Emlen at Cornell University have confirmed that some species of birds can make use of the stars to choose a direction of flight.

This conclusion is based on an elegantly simple experiment. During the seasons when they migrate, caged small birds become restless at night, instead

of quietly sleeping. During this period of migratory restlessness, they will continually hop from one perch to another. Kramer and Sauer put such birds in a
circular cage with a transparent bottom; then, placing it out of doors with a clear
view of the sky, they lay underneath the cage and kept track of the direction in
which the birds hopped. They found that in the spring such caged birds tended
to spend more time on the north side of the cage and in the fall, more time on
the south side. On overcast nights, when the stars were not visible, the warblers
appeared to be completely disoriented and hopped in all directions equally. By
moving the cage indoors to a planetarium, Kramer and Sauer were able to show
that the birds would still orient to the north in the spring, but if the star pattern
was reversed and the northern constellations appeared in the south, the birds
would also reverse their orientation, following the stars.

Several other people tried to repeat these experiments, some of them using
complex cages with perches equipped with switches and elaborate electronic
counters and computers, but no one was very successful until Steven Emlen tried
it with a superbly simple cage. He built a circular arena about 14 inches in diameter
with sloping sides too steep for the bird to stand on. He lined these sides with
blotting paper and equipped the bottom of the arena with an inked stamp pad.
When he put indigo buntings in this cage, they first stood on the bottom getting
their feet well inked and then, as they became restless, they fluttered against the
sloping sides of the arena, leaving little inky footprints. All Emlen had to do was
cover the cages with a screen or a piece of plastic, put them out under the night
sky, and then look for the direction of the footprints. A remarkable improvement
over spending the night on one's back counting bird hops—and no computers
needed!

Using this technique, Emlen has confirmed Kramer and Sauer's major point:
that indigo buntings, at least, can use the stars to choose a direction. But migration
involves more than simply flying in a specific direction. Many times birds are
blown off course by crosswinds, and if they are to return to their original breeding
ground, they must compensate for this displacement. The bird must first determine
where it is in relation to its goal, in this case, its breeding ground. Having
determined this, it can now choose a new compass course to get there. This
concept of a two-step system was described by Kramer as a "map and compass"
system; the "map" was the process of locating oneself relative to the goal and
the "compass" was the directional cue, like the familiar magnetic compass, used
to keep to the chosen course.

Put yourself in the position of the bird. You suddenly find yourself over an
expanse of open water. Your compass tells you which way north, south, east,

and west are, but before that will do you any good, you have to know where you are in relation to where you're going. Is it the Atlantic, Pacific, or Indian Ocean? Or is it a great lake? This question has to be solved before a compass can help. This process,which Kramer called map, is also known as true navigation. Once you have determined that home is west, a compass is very useful for flying in that direction.

The bird that has been blown off course must then have some way of navigating and thereby correcting its compass course for home. The obvious suggestion was that this process was also accomplished using the stars. This idea is easy to test in a planetarium: one can simply display the star pattern that would be seen from a place east or west of the bird's normal migration route at any given time and note if the bird alters its heading or the direction of its inky footprints to compensate for the apparent displacement.

Sauer reports that this is exactly what happened in an experiment using European warblers; the birds did, in fact, alter their compass heading to correct for the apparent displacement. Emlen, however, has not been able to show this in indigo buntings. The buntings keep their same headings no matter how far east or west the sky pattern has been moved. It seems, then, that at least some species of night-migrating birds can and do use the stars as a compass; whether they can also use them as a map to navigate by is less clear.

But even if birds do use the stars, this cannot be the whole story. A lot of migration occurs on cloudy nights. And by using radar it has been possible to follow flocks of migrating birds even under overcast conditions. Such studies have shown that birds are well oriented under an overcast, and some recent observations suggest that birds are oriented even on the rare occasions when they fly within the clouds themselves. Clearly the birds must have alternative methods of navigation, and when one cue is taken away they must be able to switch to some other strategy. It is probably a serious mistake, therefore, to search for *the* way in which birds navigate; most likely there are many different sources of navigational information. It would be nice, though, to know what they are using to find their way in the middle of the night, inside a cloud.

Many problems arise in working on migratory birds. In the first place, migration only occurs for a month or two in the spring and fall. This means a frantic period of investigation, then a long wait until the next season. Secondly, it is hard to do experiments on freely migrating birds. William Cochran at the University of Illinois equips thrushes with small radio beacons and then waits, often for several days, for the radio-equipped birds to decide to migrate. Emlen has found he can often persuade birds to migrate by sending them up in balloons and releasing them at several thousand feet. Short of such ingenious and heroic procedures, migratory birds are hard to work with. What one would like is the white rat of the bird world, and to a large extent the homing pigeon fits that role.

The homing, carrier, or racing pigeon—a relative of the common street pigeon—has been bred for generations to return rapidly to its home loft when released away from home. Today they are still raced for sport. In a major pigeon race the fastest bird home may win several hundred dollars for its owner; not surprisingly, there is intense competition to breed and train the fastest possible birds. From the standpoint of understanding bird navigation, homing pigeons are almost ideal. They work all year long; they return to the investigator's loft, so that each bird can be used many times; they are easy to handle and are nice and big, so that one can attach all sorts of instruments and the birds can still fly. Their only drawback is that they don't migrate.

At the same time, homing pigeons routinely do what a migratory bird that has been blown off course does—a pigeon taken from its loft and released at some place where it has never been before will usually circle once or twice and then head for home. It must, therefore, have some way of determining the direction to the home loft. It must be able to navigate. This ability has attracted many investigators, and the literature of experiments on pigeon navigation is huge. In spite of this enormous amount of work, when Martin Michener and I began our research in 1962, we found that little information existed about what an individual pigeon does between the time it is released and the time it arrives at the home loft. Flocks of pigeons had been followed by airplanes, but because the pigeon is a highly social bird, it is reluctant to fly alone, and the path of a flock must represent some consensus of the orientation of its members.

Michener's idea was to follow individual birds on their homing flights, and to do this he built small radio transmitters that the pigeons could carry. My part of the scheme was to learn to fly an airplane so that we could follow a transmitter-equipped pigeon wherever it might roam. We had our share of problems in getting airplane, radio, trackers, and pigeons all to cooperate. On one of our first flights, all went well until the pigeon decided it had flown enough and sat down on the telemetry antennas at MIT's Lincoln Laboratories. Our transmitters are very low power, but placed directly on an ultrasensitive antenna, they become rather strong. Lincoln Laboratories, however, is a restricted area so we could not get close enough to throw a rock at the pigeon to persuade it to move on; all we could do was wait. When it eventually flew off, we dashed for the airplane and managed to track it toward the loft, whereupon it promptly landed again—this time on the grounds of a mental hospital. After some discussion we agreed that it didn't seem wise to wander around the hospital ground with our portable radio-direction finder—we could easily imagine the discussion with a guard if we said we were looking for our pigeon.

Actually, the pigeon-tracking radio system worked remarkably well—we could follow a single bird for days if need be. Frequently we would leave a bird roosting at night and find it in the same place before daybreak the next morning. In this way we learned a great deal about the behavior of individual pigeons. One of the first things we learned was that pigeons are either not familiar with the federal air traffic regulations or they choose to ignore them. Our birds were just as ready to fly over Boston's Logan Airport as anywhere else. When we radioed the radar approach control that we wanted to enter their control zone to follow a pigeon, there was usually a long moment of silence while the controller no doubt wondered if somebody was pulling his leg or whether there really was some nut up there following a pigeon.

While following birds that were released from places where they had been many times before, we found they took similar but not identical routes home each time. They were not simply following a set sequence of landmarks to the loft. Furthermore, their track was no more irregular on days with poor visibility than on days when you could see for miles. Nor were the birds blown off course by crosswinds. In fact, they seemed bothered only when the sun was obscured. Then they would stop flying and sit, frequently for hours, while we circled overhead in the airplane.

Next we took the birds south to New Bedford, Massachusetts, a place they had never been before. If the pigeons were using landmarks, we expected that when they were released in a new place they would search for something familiar. If they were simply using a compass and flying in the direction that had gotten them home from their training point west of the loft, they should fly east. But if they were really navigating, they should fly directly home, in this case, north. We found that all three things occurred; pigeons often would start out in their trained compass direction, then switch to navigation and head for the loft, and finally appear to use landmarks to find the loft itself. But there was great individual variation in strategies—some pigeons always flew an initial compass course; others never did.

One of our birds, B77 (a blue band, no. 77), always flew directly home. Another bird, WY, was trained along a route that had a prominent hill on the way home. Released in a new place, it invariably flew to the nearest mountain, then turned and headed for home.

The point is simply that pigeons are individuals, and they clearly use a variety of strategies. Following individual birds gave us an insight into what each bird was likely to do; then when we gave the same birds some experimental treatment, we could recognize any changed behavior.

Our first idea was that the birds were using the sun. It had been known for a number of years that pigeons can use the sun as a compass and can compensate for the sun's apparent movement through the sky by using an internal clock. But two Englishmen, G. V. T. Matthews and C. J. Pennycuick, suggested that the navigation, or map, might also be based on the sun. They proposed that a pigeon might compare the sun's position where it was released to its memory of the sun's position at the home loft at the same time of day. From this comparison, the pigeon could determine the direction to the loft. This was an appealing idea not only because it is somewhat similar to what human navigators do but also because it made specific predictions that could be tested. For example, a pigeon that had been released north of its loft would see a sun at noon that was lower in the sky than the home sun at noon. Similarly, for a pigeon released south of the loft, the noon sun would be higher in the sky than at the home loft. Of course, because pigeons navigate at all times of day, not just at noon, they would have to have a way of estimating the sun's noon altitude at any time of day, but let's ignore this complication for the moment.

To test this idea, Michener and I arranged a small cage in a corner of the loft. From this cage the only view of the outside world was through a window, the bottom half of which gave a view of the field in front of the loft and of the lower part of the sky. The top portion of the window showed the pigeon the sky's upper region and allowed it a view of the sun for about three hours centered around noon every day. We hoped the pigeons did not notice that their view of this top part of the sky and sun was seen through a pair of mirrors. By adjusting the angle of one of the mirrors, we could make the sun appear either too low or too high in the sky. In short, we could make it appear to a pigeon, attentive to the sun's altitude at noon, that the loft had been moved either north or south.

We exposed pigeons to this "mirror box" for ten days to two weeks, then released them in a place they had never been before. If they "believed" what the mirrors had shown them, we expected them to fly toward the place where the sun really behaved as it appeared to in the mirror box.

Unfortunately, the pigeons simply flew home.

Then we wondered if the pigeons had homed normally because the mirrors had not, for some reason, fooled them. This being possible, we switched to a more radical form of treatment: we began to upset the pigeon's internal clocks. A bird that is using the sun for navigation must have an accurate clock to determine whether it has been displaced east or west of home. If a pigeon determines north-south displacement by noon sun altitude, it might use what is, in effect, a comparison of sunrise times to determine its east-west position. Suppose a pigeon notices that the sun rises at 6:00 A.M. in its home loft in Lincoln, Massachusetts. Now the pigeon is displaced and sees a sun that is not as far along its path, indicating a sunrise time of 7:00 A.M. Obviously, such a bird is west of the loft, and must fly east to get home.

The bird could use the time when the sun reaches its highest point in the sky, the noon position, in the same way. The time of sunrise or of local noon varies about four minutes for every fifty miles you travel east or west. If you go from Boston to Albany you will notice that both the sunrise and the local noon are eight minutes later in Albany. Normally, we ignore such differences until they reach about an hour; then we call them time zones. They are about 750 miles wide.

A number of people have shown that if a pigeon is put in an artificial environment where the lights go off and on out of phase with the real day, its internal clock will gradually shift to the new time schedule. Thus, if sunrise and sunset are 6:00 A.M. and 8:00 P.M., respectively, but the lights in the box go on at 7:00 A.M. and off at 9:00 P.M., the pigeon has experienced a normal day length, but is shifted one hour out of phase—its internal clock is one hour behind local time. And there is, in fact, a place where the sun altitude is the same as at the loft, but the local sun time is exactly one hour behind. Such a place would be 750 miles west of Lincoln, Massachusetts, or near Chicago. If a pigeon were navigating by the sun, it might well home to this place rather than to the loft.

To test this idea we used pigeons with clock shifts of 0, 5, 10, 15, 20, 60, 120, and 360 minutes, and then released and tracked them from places they had never been before. All except the birds with the 360-minute (six-hour) shift homed normally. The birds with the six-hour shift headed in every direction.

That birds with a six-hour clock shift show poor homeward orientation is not surprising when you remember that pigeons also use the sun as a compass. Since the sun moves 15 degrees an hour, a bird with a six-hour error in its clock should make a compass error of 90 degrees. Klaus Schmidt-Koenig in Germany has shown that this is exactly what happens. Suppose you want to fly east. The real time is sunrise, 6:00 A.M. Your internal clock says that the time is six hours later, noon. To fly east at 6:00 A.M., you fly roughly toward the sun, but because your clock tells you it is really noon, you know the sun is in the south and that to fly east, you must fly 90 degrees to the left of the sun. And this is exactly what the birds do, although presumably they do not go through the reasoning process I have described.

Our clock-shift experiments seem to have affected the birds' sun compass, but to have had absolutely no effect on their navigation. In desperation we considered the possibility suggested by Klaus Hoffmann—that the birds might have two clocks. One of these would be a "sun compass clock," easily set to local time; the other a "navigation clock," highly resistant to resetting.

This idea led us to perform a really radical experiment. It has been found that heavy water, in which the hydrogen atoms have extra protons in their nuclei, slows down all biological clocks on which it has been tested. Lee Snyder found that 30 percent heavy water in drinking water slowed down the activity rhythms of pigeons by about 6 percent. We took a group of experienced pigeons, put them in a clock-shift box where the lights came on and went off at irregular intervals, and gave them 30 percent heavy water to drink for about two weeks. At the end of this time, their internal clocks should have been thoroughly scrambled. But we could not just take these birds out and release and track them. Their sun compass clocks would also have been upset and it would be impossible to say whether our treatment had affected their map or only their compass.

Instead, we transported the birds in a closed container to Ithaca, New York, about 250 miles west of Boston. William Keeton and Andre Gobert held the birds for us in an open cage where they had an unobstructed view of the sun. After ten days to two weeks in Ithaca, drinking normal water and with a view of the sun and a normal day, the pigeons should have reset their clocks to Ithaca time. Next we took all the pigeons to a spot between Ithaca and the home loft. If the pigeons' navigation clocks were set to Ithaca time and they were using the sun, they should fly west to Ithaca. If they were simply confused, which seemed the most likely possibility, they might fly in any direction. But if they weren't bothered by our treatment, they would fly home.

And that is exactly what they did. Their homeward bearings and tracks were even more accurate than those of pigeons just taken directly out of the loft. This result, together with Keeton's finding that his pigeons showed excellent orientation even under total overcast, leads us to believe that the sun is probably not the basis of pigeon navigation. At the very least, there is some other series of cues that pigeons can use to get home.

It occurred to us that the pigeons might be getting some information during the trip from the loft to the release site. Such a possibility seemed unlikely, but to test it we anesthetized the birds at the loft and drove them to the release point fully anesthetized, in a covered cage, and in a strong, fluctuating magnetic field. Such an experiment may seem to have too many variables, and indeed, that is the case. But if all of these treatments have no effect, then one has ruled out several possibilities at once. And that is exactly what happened: after the birds had recovered from their anesthesia, they homed just as well as the controls. This result indicates that the pigeons are probably obtaining their navigational information at the release site rather than during the trip out.

There is other evidence suggesting the same thing—each release site seems to have its own, characteristic error. That is, birds released at a particular point may not head precisely toward home. They may instead choose a direction as much as 90 percent from the correct home bearing. This error is consistent both from day to day and even for pigeons from different lofts.

Whatever it is the pigeons are using probably does not depend upon their ability to see. In a series of very exciting experiments, Hans Schlichte and Klaus Schmidt-Koenig have shown that pigeons can home even when their eyes are covered with translucent contact lenses, which prevent them from seeing anything but a general, diffuse image of very close, large objects, rather like the view one gets of an approaching person through a ground-glass door. Because pigeons equipped with these lenses showed homeward orientation as accurate as that of birds equipped with transparent lenses, it seems unlikely that vision plays an essential role in pigeon navigation.

But if we eliminate vision, what is left? An obvious idea is that pigeons might somehow use some aspect of the earth's magnetic field. This is a very old idea, and the experiments of Henry L. Yeagley, a physicist at Dickinson College, for example, show no convincing effect of equipping pigeons with small magnets. Once again it was the work of Keeton at Cornell that showed that, under overcast conditions when the sun was not visible, pigeons equipped with small magnets showed poorer orientation to the loft than did birds carrying equivalent brass weights. Under sunny conditions there appears to be no effect when the birds are released close to the loft, but if they are released at greater than usual distances, the magnet-equipped birds are frequently disoriented, whereas the birds with brass weights are not.

With our birds I found no differences when the pigeons were released at 50 to 70 miles, but on two releases from 120 miles on sunny days, the magnet birds were completely disoriented, the brasses were not. There are, however, puzzling aspects to these experiments. For one thing, the results are variable; one day there will be a clear difference between the two groups; the next day there will be none. Secondly, the pigeons with magnets do not all seem to fly on a consistent bearing— they simply fly off in all directions.

It occurred to us that the field of a small bar magnet is restricted and, as the pigeon moved its head, the magnetic field around the head would vary greatly. Of course, there is no particular reason to believe that a pigeon's magnetic receptor, if it has one, would be located in the head. It might, for all we know, be in the left hind toe. But the head seemed a likely place to begin.

We equipped the pigeons with a pair of small coils and a battery. One coil was glued to the top of the pigeon's head like a hat, the other served as a collar around its neck. By changing the amount of current through the coils we could accurately vary the strength of the magnetic field around the head. We began with very weak fields, only 0.1 gauss (approximately 1/6 the strength of the earth's normal field of 0.6 gauss). Control birds carried the same apparatus, but the coil was not connected to the battery. The effect of this procedure was to increase the scatter of the directions that the experimental birds took, as compared to the controls. It appeared that this effect is relatively consistent: on 14 out of 19 releases the experimental birds were less well oriented than the controls.

Most recently we have had a hint of an even more impressive effect. Robert Green, an undergraduate at the State University of New York at Stony Brook, made two releases of birds with coils designed to produce somewhat stronger magnetic fields (0.6 gauss), under overcast. His experimental and control birds were identically equipped with coils, except that ten birds had their batteries connected one way and eight had the battery connections reversed. The magnetic field of the coils was thus oriented one way for the first group and in the opposite direction for the second. That is, the north pole of the induced field was toward the bird's head in one set and toward its tail in the other.

When the two groups of birds were released and tracked from New York City, fifty miles west of the loft at Stony Brook, Green found that one group of birds flew directly east toward the home loft but that the other birds flew in an almost exactly opposite direction! While the number of birds used and the number of releases are not great enough to verify these results, they are exciting. If these experiments can be repeated, it suggests that an applied magnetic field can drastically upset a pigeon's orientation and that the effect of the field depends on its polarity. If by altering the direction of the applied magnetic field, we can make pigeons fly in predictable directions rather than simply flying home to the loft, it indicates that magnetic fields are in some way related to pigeon navigation. After repeated experiments, the differences probably will disappear—that seems to be one of the problems in working with pigeons. But they just might not, and it is that hope that keeps us going.

Suppose that we are successful and find that artificial magnetic fields do have an effect: Have we then found the basis of the pigeon's navigation? I suspect not. It is hard to see how a pigeon could use the earth's magnetic field to tell it where it is in relation to home. It seems much more likely that if the birds are using magnetic cues, they serve as an auxiliary compass to be used when either the sun or stars are obscured. Only further investigation will tell.

We are sure that birds use a wide variety of cues to migrate or home over the surface of the earth. Migrating warblers use the stars as a compass, and pigeons use the sun. No doubt both can use landmarks. But how they navigate is still unknown. And in a way that's rather nice. Here is a problem that many people have worked on for a great many years; we have all seen the phenomenon, it is all around us. But the birds won't tell—at least not until we ask the right questions.

6

Psychophysics and Hearing in Fish

WILLIAM N. TAVOLGA

The question of whether fish can hear seemed to be well established as long ago as 1820, when E. H. Weber, along with his excellent anatomical studies of the human ear, described the ears of fishes. He theorized that, although the fish has no external ear, the swim bladder acts in a manner analogous to the middle ear of man. That is, it receives the sound energy and transforms it into vibrations of the fluids of the inner ear. It took almost 150 years to prove Weber's contention. Around the turn of the century, some well-controlled experiments clearly demonstrated that fish could hear. Sounds of buzzers, struck objects, and all sorts of natural and artificial sounds were used, and fish were found to respond to many of them. Karl von Frisch even trained a catfish to come to him when he whistled. In the early 1900s, G. H. Parker theorized that a fish can detect sound underwater in two ways. In addition to receiving sounds by way of the swim bladder and inner ear, he said, the lateral line system is also sensitive to sounds. The lateral line consists of a series of minute sense organs embedded in pits and tubes that usually form a thin, visible, lengthwise line on each side of the body of most fish. There are also a number of interlacing tubes and separate pits on the head. This entire system was thought to be primarily sensitive to movements of water currents and low-frequency vibrations. Once it was established that fishes could hear, the next question was how well? For instance, what frequencies can a fish detect? The first of these seemed to interest most investigators, particularly the peripheral question: what is the highest frequency a fish can hear?

Most of the experiments to determine the frequency range of fish hearing were behavioral. That is, some response on the part of the entire animal was used as a criterion, although a few observations have been made in which the responsiveness of the sense organ was studied directly. In the latter case, electrodes were placed on the nerve fibers coming from a receptor and the signal was "wire tapped." This technique is extremely difficult, as the auditory nerve is short and deeply embedded in bone. Some success was achieved in sharks, whose large size and cartilaginous skulls made the technique possible. This wiretapping has also been done with the lateral line, where it is a bit easier. Such data, however, are useful only to show the potentialities of the sense organ. That is, we can tell what stimuli the sense organ can react to and what messages it sends along the nerves leading from it to the central nervous system, but we cannot know, from such information, what the animal will do with these signals. A classic example of this is one in which electrophysiological techniques have shown that the ordinary cat should be able to discriminate colors. Behaviorally, however, the cat is color-blind. It is apparent that somewhere in the central nervous system this color information is discarded. Thus, if one is primarily interested in the behavior and ecology of the organism, it is more desirable to determine what the whole animal will respond to, rather than to measure the capabilities of the sense organs.

The methods that have been used to determine auditory capacities have, for the most part, involved conditioning the animals to respond positively to a sound associated in time with the presentation of food. Another technique, utilized primarily by investigators in the Soviet Union, is that of classical conditioning. Here the fish is exposed to the test sound and this is followed shortly by a mild electric shock. A positive response is any sudden movement, involuntary reaction, or even a respiratory or heart rate change that occurs every time the sound is made.

To summarize the results of all these investigations, fishes must be separated into two groups. The majority of species has an upper frequency limit of about 2000 or 2500 cycles per second (abbreviated to c.p.s.). This is a pitch about two and a half octaves above the standard middle A on the piano. The second, and smaller, group belongs to the order Cypriniformes, and are considered the hearing "specialists." This order includes the catfishes (including bullheads), carps, minnows, characids, and gymnotid eels. It has been reported that bullheads have responded to frequencies up to 4000 c.p.s., and certain minnows may hear as high as 8000. This last is almost an octave above the highest note on a piano—a C at 4186 c.p.s. (The Cypriniformes and others show a close association between the swim bladder and the inner ear, which will be discussed later.)

Sound intensity must also be measured. What is the minimum level at any specific frequency to which the animal will respond? Some investigators attempted to measure this, but in most cases they only tested one of a few frequencies. Only two reports I know of attempted to determine the complete hearing curve for a fish. Autrum and Poggendorf, in 1951, worked out the audiogram for the fresh-water brown bullhead, *Ictalurus nebulosus*. In this sort of graph, the sound level is given on the left side (ordinate) and the frequency along the baseline (abscissa). The lower the point on the curve, the lower the threshold, that is, the greater the sensitivity at that frequency. In 1961, Kritzler and Wood made a similar audiogram of the bull shark, *Carcharhinus leucas,* at the Lerner Marine Laboratory at Bimini, Bahamas.

In practice, sound intensity is measured by suspending an underwater microphone (hydrophone) in the water. Since the hydrophone transforms sound pressure into electrical pressure, the voltage output of the hydrophone is exactly proportional to the pressure of the surrounding sound field. If the hydrophone and its amplifier are properly calibrated, the sound pressure can be determined with considerable precision. Sound can be measured in terms of pressure, that is, force per unit area. The units we use are dynes per square centimeter—one dyne per square centimeter is known as a microbar. This, in turn, is approximately equal to one-millionth of average atmospheric pressure.

In human hearing (out of water, of course), the threshold at 1000 c.p.s., based on an average of many individuals with "normal" hearing, is .0002 microbar (figure 6.1C). This value is often used as a standard, and all other sound pressures are related to it. If a person is asked to discriminate one sound intensity from another, the minimum difference he can detect is defined as a decibel, but the absolute magnitude of a decibel depends on where one starts. At a low sound level, a decibel is much smaller than it would be at a high level. The decibel scale is a logarithmic one that is based upon an equation in which it is assumed that human hearing, and that of all other animals, follows a logarithmic law. Although most evidence indicates that human hearing follows some other type of equation, and that there is no reliable evidence for any other species, we stick to this decibel scale and use it in acoustics, electronics, and many other fields because it is convenient. We can decide, for example, to choose the .0002 microbar value as a reference value. This would then equal 0 decibels—as in the human audiogram.

In many phases of acoustics, especially in underwater work, the 1 microbar value, rather than the .0002, is taken as the reference level of 0 decibels. This is actually a more objective reference and has come into wider usage in recent years. Conversion from one reference level to another is a simple matter of adding or subtracting 74.

Water is much more resistant to the propagation of sound than is air. This means that to produce the same effect, sound pressure in water must be much greater. Conversely, at the same pressure, the acoustic energy in air is greater than in water. Sound volume can be expressed in two ways. The usual, and more convenient, way is in terms of pressure in decibels with reference to some standard pressure value such as 1 microbar. However, we can also express acoustic energy in terms of intensity or power. This is normally given in watts per square centimeter.

In air, the human hearing threshold is .0002 microbar at 1000 c.p.s. This can also be expressed in acoustic power as some fraction of a watt/cm². It happens to be one ten-quadrillionth of a watt, more simply written as 10^{-16} watts/cm². Actually, we are primarily concerned with this power figure, since it is the energy of the sound wave that we receive. Pressure is a more convenient measure to use, but we must insert a correction if we compare acoustic pressure in air with that in water. This correction is approximately 36 decibels. That is, .0002 microbar in air is actually 36 decibels (of power) higher than .0002 microbar in water. To put it another way, given the same power, the pressure in air is 36 decibels lower than in water. All this is because of the higher density and incompressibility of water. Because most measurements are made in pressures, we now have to convert all our figures into equivalent power units if we are to make a proper comparison of sound in air and in water. Such a comparison is shown below.

In an attempt to answer the question of how well a fish hears, I collaborated with Dr. Jerome Wodinsky, a psychologist at Brandeis University, to find a conditioning experiment that would allow a fish to give a reliable, repeatable, and unequivocal answer to a question. The simplest answers are, of course, "yes" or "no." A "maybe" cannot be tolerated. (An animal must be placed in a situation in which it has only to say "yes." It need do nothing to say "no.") This sort of limitation is particularly important in sensory studies. As the stimulus approaches its lowest detectable level, the subject, be it human or fish, becomes unsure of whether he detects it or not, and begins to try to say "maybe."

The objective technique we used is called "avoidance-conditioning." It was first demonstrated in dogs by the famous Russian psychologist I. P. Pavlov. At the sound of a bell or the flash of a light, the dog had to lift its forepaw. If it did not do so, it would receive a mild, but annoying, electric shock. By raising the paw immediately upon the presentation of the sound or light, the animal "avoids" being shocked. This is a potent form of conditioning, and is retained for long periods. It also forces a clear, unambiguous response from the subject.

Most theoreticians now agree that the acquisition of this avoidance response takes place in two stages. First, the animal learns to make the response that will turn off the noxious stimulus. This has been variously called classical, or Pavlovian, conditioning. The animal, therefore, learns to *escape* from the noxious stimulus. In the second stage it learns that the sound precedes the shock and that the same escape response can be used to *avoid* the shock.

In applying this method to the study of hearing in fish, we used a "shuttle box." This is an aquarium with two compartments separated by a shallow barrier. The water level is adjusted so that the fish can swim from one side to the other, yet will not remain on the barrier, because the water is too shallow. The sound source is concealed beneath the center barrier, and the entire tank is shock-mounted and insulated to reduce the noise level inside, reduce reverberations, and prevent the animal from seeing anything that might serve as an additional cue. The procedure is to turn on the sound and, after a predetermined period of five or ten seconds, administer a series of short, intermittent electric shocks. The fish first learns to escape the shock by crossing the barrier, because as soon as it does so, both shock and sound are stopped. This phase takes only a few trials. Each time the fish must move from one compartment into the other and can start from either for the next trial. The spacing of the trials must be varied, or the fish learns the length of the intertrial interval and begins to anticipate the shock.

The second stage of learning takes a little longer. In most species, three to six days of 25 trials a day are required before the subject begins to avoid regularly. A positive response, then, is one in which the fish swims across the barrier as soon as the sound goes on, but before it receives a shock. The response eventually becomes extremely reliable—so much so that the shock administration becomes unnecessary.

Once the avoidance-conditioning was well established, we changed the sound level. Generally we started with a pure tone—a single frequency—at an intensity we felt sure the fish could hear. After each avoidance, the sound level was lowered in steps of 2 or 5 decibels, so that the intensity would be lower at the next trial. This was continued until the animal missed—that is, did not avoid, but received the shock and escaped. This was recorded as a "no" answer. After each "no" the sound level was raised for the next trial. When the results are plotted on a graph, a zigzag line stretching across the paper is produced. If the tops of the "zigs" and bottoms of the "zags" are averaged, we can calculate the threshold for that frequency. It must be remembered that a sensory threshold is not an all-or-none situation, and there is a degree of probability that we will get some "yes" answers below the threshold and some "no" answers above it. A "threshold" is a stimulus level the subject can detect and respond to 50 per cent of the time, and is thus a statistical value, not an absolute one.

It is necessary to repeat such determinations a number of times using different subjects, so that the value obtained for the given frequency is more reliable.

Eventually, this is repeated at different frequencies, and an audiogram for the species can be plotted.

Other species have given us similar curves, but as many as 20 decibels higher or lower. So far we have worked out these audiograms for nine species of marine fish. They represent a large majority of salt-water fish, although none is a so-called specialist in hearing.

In order to make the determination less subject to human error and bias, the equipment for this study has been partially automated, to allow us to graduate from working with a pair of hand-operated switches and watching the movements of the fish by means of a mirror. The observer sits before a control panel and pushes a button. This button automatically starts and continues a trial. The sound goes on and, if the fish does not avoid, the shock continues according to a preset schedule. When the fish crosses the barrier, a beam to a photoelectric cell is broken and the sound and shock are automatically turned off. A clock is also part of this apparatus, so that the time it takes a subject to respond is recorded. In addition, a counter keeps track of the number of times the animal crosses the barrier during the intertrial interval. These data are important because we want to be informed of the activity of the animal—how often it crosses the barrier and if these intertrial crossings represent "false alarm" responses. All this is multiplied by six in our apparatus, so that we can observe and test six animals in six different tanks simultaneously. Eventually, we may have to feed our data into a computer so that all the calculations and analyses can be performed on a large number of figures in a short time.

At this point, we can begin to make some generalizations as to what fish— at least marine fish—can hear. For most species, the upper limit is about 1500 to 2000 c.p.s., which is about one and a half to two octaves above middle A. Above this point, the sound levels become so high that they may actually cause the animal physical discomfort or pain. The most sensitive range is from about 200 to 800 c.p.s., or a little more than the center octave on a piano. In this range, the sensitivity of some species comes close to that of the human ear, but we must remember that we are comparing a fish hearing in water to the human ear in air, and this may not be a fair or meaningful comparison. The lower frequency limits are difficult to set, because it becomes a matter of definition as to how low we can go and still call it "sound." Many fish seem to be at least as sensitive to a 20 c.p.s. sound as we are, but sound under water presents a special situation, because water is much denser than air, and is not easily compressible. This density and incompressibility offer resistance to the flow of acoustic energy, and although the transmission may be more efficient—the velocity of sound in air is about 1080 feet (330 meters) per second, while in sea water it is about 4900 feet (1500 meters) per second—the amount of energy required to propagate a sound in water is almost 150,000 times greater than in air.

Because of this, another factor becomes important—the actual particle displacement that results from the vibration of a sound source. This displacement—called the near-field effect—is of considerable significance at low frequencies, and at a short range from the source. Close to the sound source, therefore, the acoustic energy is in two forms: one is the pressure wave (as exists in airborne sound) and the other is an actual physical vibration of the water itself. Which is it that the fish receives? We can safely say that at frequencies above 800 c.p.s. the fish can respond only to the pressure wave, and at lower frequencies and at distances of 20 or 30 feet or more, the pressure wave is still paramount. In the range of the near-field, however, the displacement effect is probably most important.

Even when dealing with a pure far-field pressure phenomenon, however, we still get into complications. If a bubble of air is placed in the path of a pressure wave, the bubble will vibrate and produce a near-field effect in its vicinity. Two scientists at the Bell Telephone Laboratories, G. G. Harris and W. A. van Bergeijk, proposed that the swim bladder of a fish may act in such a manner. The inner ear, then, would receive this local near-field effect. In addition, there are the complications of all the reflections and reverberations that can take place under water. Not only is 99.9 percent of sound energy reflected back from the water surface, but layers of water at different temperatures can serve as sound mirrors. These factors become exaggerated in the small aquariums in which we test the fish's hearing. All we can say at this time is that we can obtain thresholds for some form of acoustic energy, but cannot determine exactly in what form that energy is received.

Now let us approach the problem of *how* a fish hears. Compared to the human ear, that of the fish appears simple. This is deceptive. The fish does not have a helical cochlea. Rather, the inner ear is a sac of fluid, with areas of hair cells protruding into a liquid (endolymph) in which float one large and two smaller bones. Movements of the ear bones (otoliths) and liquid stimulate the hair cells, and signals are sent along the auditory nerve to the brain. Thanks to the brilliant work of von Kekesy, we know something about how our cochlea operates to discriminate one frequency from another, but there is nothing comparable in the fish ear. How does a fish discriminate pitch—or does it? Some studies on the goldfish indicate they may, but the evidence is not clear as to whether there is a true frequency discrimination or if the apparent discrimination is actually based on intensity differences.

How does the acoustic stimulus reach the inner ear? In the hearing specialists, like the catfish, there is a series of four pairs of small bones leading from the swim bladder to the inner ear fluids. Experiments have shown that damage to these bones reduces the hearing capacity. The bones and their probable functions were first described by Weber, and he proposed that they act in a manner analogous to human middle ear bones in transmitting air vibrations to the endolymphatic fluids. These ossicles have since been named the Weberian apparatus. As mentioned before, the swim bladder—even in fishes without the Weberian apparatus—can function as a middle ear by creating a local near-field effect. It is quite possible for sound vibrations to reach the inner ear directly by way of bone conduction through the skull. Sharks do not have a swim bladder, but it can be shown that they have as good hearing as some bony fishes with swim bladders. In our own work, differences in the sensitivity of marine fishes cannot be correlated with size and location of the swim bladder.

The swim bladder of fishes has a number of functions. In most cases it serves as a hydrostatic organ—that is, the buoyancy of this bubble of air counteracts the tendency of the fish to sink. By changing the volume of the bladder, the fish can change its own buoyancy. In some cases, the bladder is used as a temporary reservoir of oxygen, and in a few forms it even acts as a lung for breathing air directly. In many species, the swim bladder acts as a loudspeaker for sound production. As a hearing organ, it is undoubtedly important, because the body of a fish is almost transparent to water-borne sound. The bladder, therefore, can act as both a loudspeaker and a microphone. If we project pulses of sound, as in sonar, we can locate fish by the reflections of the sound pulses. Most of this reflected sound comes from the swim bladders, and very little from the rest of the fish's body. The swim bladder, therefore, serves as an acoustical discontinuity, and presumably is of prime importance as a sound detector.

We must not neglect the function of the lateral line system in sound detection. The structure of the system's individual sense organs is ideally suited for the detection of movements of water. Indeed, it was shown by a Dutch scientist, Sven Dijkgraaf, that the lateral line can give the fish information about water currents and moving objects, and can even be used to locate the position of obstacles in complete darkness. As underwater sound produces a significant displacement at close range to the sound source, that is, the near-field effect, this, too, can be received by the lateral line. Therefore, at close range and at low frequencies, the lateral line is also a hearing organ.

In this respect, the lateral line has certain advantages over the ear. Sound pressure, as such, is not directional. In humans, if one ear is plugged, it is impossible to determine the direction from which a sound comes. By using both ears, directionalization is possible, because of the different times it takes sound to arrive at each ear. In essence, the fish has only one ear, because the spacing between the two receptors is so small and the speed of sound is so high. In the near-field, however, the displacement energy *is* directional, and the lateral line organs are dispersed widely on the animal's body. Harris and van Bergeijk propose that the fish can locate the sound source, but only within the limitations of the near-field.

It is clear, then, that fishes can respond to subsonic vibrations of the water, and to sonic vibrations up to at least 2000 c.p.s., with some specialists able to perceive up to 8000 c.p.s. The most sensitive range is below 800 c.p.s., and here many species appear to have a sensitivity comparable to that of the human ear. The swim bladder is the main sound receiver, transmitting its vibrations to the inner ear, but the lateral line system is also a hearing organ. The latter is particularly sensitive in the low-frequency and subsonic range, and at short distances it can locate sound sources. Such conclusions are based on cooperation among psychologists, physicists, and biologists.

7

Invertebrate Learning

MARTIN J. WELLS

Colorful, graceful, and intelligent, *Octopus vulgaris* Lamarck is a beautiful animal. It will live well in a covered aquarium, given only a liberal supply of clean, well-aerated sea water, plenty to eat, and a heap of stones or bricks to create a "home" into which it can retire between explorations around its tank. Once it has settled down, the animal will spend most of its time sitting in the home, seeming to watch with interest everything that is happening in and around its tank.

Octopuses appear never to sleep. Anything that moves in the tank at once attracts an animal's attention. If it is smaller than the octopus, it will almost certainly be attacked; the animal emerges from its shelter to stalk the intruder, and finally leaps upon it, covering the prey with the web that joins the eight arms. The food is carried home to be eaten.

These animals have voracious appetites—a small octopus will double its weight in a month, if given the chance—and in aquariums they quickly become tame. They even come out of their homes to "greet" people once they have learned that, usually, people mean food. Their preferred diet is crabs, but they will also eat other crustaceans, bivalves, and pieces of fish. Little octopuses are eaten by big ones, and, in general, the animals are antisocial, predatory, and better kept apart in individual tanks.

As experimental animals they are of great interest because of the ease with which they can be taught to make a wide range of visual, tactile, and taste discriminations. If an octopus is given food as a reward when it grasps one of two objects that are exhibited successively at the far end of the tank, and a small electric shock (by means of a probe, with which the animal can be touched under water) when it grasps the other, the animal learns in only a few trials to attack the one and ignore the other. In this way, it can be taught to recognize figures of different shapes—squares and circles, crosses and triangles—and to distinguish among different sizes of the same shape. It can also learn, for example, to distinguish between a square and a diamond—the same shape rotated through 45 degrees—or between the same rectangle shown vertically and horizontally. Octopuses make many such distinctions rapidly, within a matter of twenty or thirty trials (individuals vary somewhat, as one might expect with any sample of intelligent animals taken from a wild population), and once having learned, remember for at least some weeks.

That the animals can so readily be taught opens up a number of opportunities. For one thing, it is possible to trace those parts of the nervous system that are concerned with learning by removing parts of the brain. Similar studies have, of course, been conducted on many kinds of vertebrate animals, and through the octopus work we are now beginning to be able to compare structure and function in the brains of cephalopods and vertebrates—two quite independent groups of animals—with results that should ultimately tell us much about the organization of both. Alternatively, by considering the problems that an octopus can and cannot solve, one can investigate some of the ways in which the animals classify the various objects they see or touch.

This second approach to brain function has formed the basis of a series of experiments conducted during the last few summers at the marine biological station in Naples. In these, octopuses, blinded by cutting the optic nerves, have been taught to distinguish by touch between objects they grasp with the suckers on the undersides of the eight arms. The initial purpose of the experiments was to provide a basis for brain lesion work on touch learning, by establishing some simple discriminations that the animals could carry out reliably before surgical interference with the central nervous system. But, as so often in research, the problem rapidly developed a series of sidelines that were interesting in their own right. One of these was the discovery of a distinct range of tactile discriminations that octopuses are apparently unable to make. These failures are particularly interesting in view of the many other tactile discriminations that are made with difficulty—not only because they tell something about the organization of the octopus itself but also because they reveal, by implication, a general truth about the organization of the nervous systems of invertebrate animals. It is the failures, their implications, and the experiments arising out of them that will be discussed in this article.

First, however, let us examine some of the tactile discriminations that octopuses can be trained to make. They can, for example, learn to discriminate among a variety of types of cylinders, although (and here at once answers are found that at first seem curious by our standards) octopuses apparently make these discriminations entirely on a basis of the texture, or roughness, of the objects concerned. They do not, one finds, distinguish between cylinders such as *A*, *B*, and *C* that are alike in this respect, and they seem quite unable to detect that these equally rough objects differ in other ways. Differences in the orientation and/or the pattern of the grooves cut into the cylinders, which we can detect readily, apparently pass undetected by them.

Octopuses can also be taught to distinguish between cylinders of different size. But once again the results are a little odd by vertebrate standards. For instance, octopuses treat cylinders made of bundles of rods stuck together as being of the same size as their component rods, rather than in relation to their overall diameter. The animal evidently classifies the cylinders on the basis of the distortion imposed on individual suckers, rather than on the degree to which its arms are bent around the objects.

As a check on this, octopuses trained to accept cylinders of large diameter and to reject small ones were offered a series of rough- and smooth-textured objects in a series of transfer tests; they accepted the smooth objects and rejected the rough. This implies that texture is equated with diameter, presumably because both are measured by the octopus in terms of the distortion of its suckers.

As with the textural discriminations already discussed, what happens at the level of the individual suckers seems to be all-important—size, like texture, is apparently determined from the degree of distortion caused by the contact. Indeed, it is arguable that size is a textural problem so far as octopus touch sense is concerned.

In addition to detecting some of the physical differences between the objects they touch, octopuses can distinguish tastes by using only the suckers. They can be trained to recognize and take or reject spongy objects soaked in solutions of such substances as quinine, sugar, or acids at dilutions well below the range of the most sensitive human tongue. In the wild, this chemotactile sense is probably of the greatest importance in their search for food, much of which will be found hiding in crevices into which the octopus can reach but cannot see. At present, we know only that the sense is exceedingly acute and that the animal readily learns to distinguish among tastes under laboratory conditions; the range of distinguishable tastes, and the way the octopus classifies them, have so far not been investigated, largely because we ourselves find it difficult or impossible to classify such chemical stimuli.

Because octopuses can learn a great many distinctions rapidly, it was somewhat of a surprise to discover that there were certain categories of apparently simple discrimination that they never seemed to learn at all. Weight is one of these. Octopuses will pick up and handle heavy objects, but seem incapable of learning that they are more than usually hard to support.

As a result, they never learn to distinguish between light and heavy objects in training experiments. They cannot be trained to pass one (say, the lighter) of two objects under the web to the mouth and to reject the other. Such results are the more curious because the animal makes obvious reflex responses to weight differences. An observer, watching an octopus being trained, can tell at once which of two objects, light or heavy, the animal is handling; the animal is muscular, and one can see the arm muscles tightening to take the strain. But the octopus itself never seems to remember the relationship between the force needed and the reward or punishment that follows when the object is passed under the interbrachial web to the mouth. Long training fails to improve performance, and the animals continue to respond at random for hundreds of trials, even when one of the objects is several times heavier than the other.

A further limitation of the animal's abilities is revealed when attempts are made to train it to distinguish by touch between objects of different shape. Training experiments show, for example, that octopuses cannot learn to recognize the difference between a cube and a sphere, although eventually, in most instances, they achieve a proportion of correct responses by treating the matter (once again) as a textural problem. We know that they make the identifications by texture, rather than by shape, because of the way they behave in training experiments in which other shapes are presented. If an octopus that has learned (albeit after long training) to discriminate between a cube and a sphere is presented with a narrow rod, it almost invariably treats this as if it were the cube, taking or rejecting it without hesitation, just as if no change in the objects had been made. Indeed, for some of the animals the rod seems to be a "better" cube than the original; they make fewer mistakes in distinguishing it from the sphere. The implication—as in the size discrimination experiments—is that the distortion of the ten or twenty suckers in contact with the object is what is pertinent, rather than the overall shape of the object grasped. A narrow rod can be regarded as a series of rounded corners; the cube has, in addition, large areas of flat surface. Therefore, the cube distorts only those suckers that happen to grasp its corners. The rod distorts all the suckers, and thus constitutes a sort of supercube, so far as the octopus is concerned, in contrast with the sphere, which cannot distort the circular rims of the suckers at all.

What have these various discrimination failures in common? Octopuses do not, it seems, distinguish the weights, the pattern of surface irregularities, or the shapes of the objects they touch. Presumably, these failures are related in some manner to the construction of their central nervous system. Presumably, also, the difficulty does not arise from any failure of the animal's memory system (since octopuses learn other discriminations so readily), but rather from a lack of some relevant sensory input to the learning parts of the animal's brain. The problem is perhaps best envisioned by thinking first about ourselves. How would we solve the problems that an octopus finds impossible?

Try examining any small object, feeling it, without looking, to determine its shape. Consider the sources of sensory information that you are using. Fingertip contact, certainly, but also information about the movements of the fingers. Without knowing the relative positions of the fingers and/or how these are moved over the surface, it is quite impossible to determine the shape of an object touched, its size, or the pattern of any irregularities in its surface.

Without knowledge of the exact position of our fingers, we can find out very little except, perhaps, something about the roughness or smoothness of the things we touch; but even there our discrimination is relatively poor. Octopuses can do better than we can because their suckers are softer and more flexible than our fingers, and their sense organs are not overlaid by thick layers of protective cells, as our fingers are. As a result, they can determine very small differences in texture, and since they are sensitive to taste as well, their tactile world has dimensions we find difficult to imagine.

To get some idea of the touch sense of an octopus, one should perhaps try examining objects with the tongue, rather than with the fingers. But with all this sensitivity, the tactile world of the octopus clearly lacks something that is present in our own. The animal seems to have no equivalent to our finger position sense. All the evidence would suggest that it simply does not know where its individual suckers are, relative to one another, or how its arms are bent around objects.

There are good reasons why this should be so. The octopus is a flexible animal. The movements of its limbs are not restricted by joints, and while this has many advantages from a locomotor point of view, it does set a major computing problem for the central nervous system. To know where my fingers are, my brain must take into account the bends at a couple of dozen different places; the problem is limited because I don't bend anywhere else. But for the unjointed octopus, it would require an elaborate computer, indeed, to assess the relative positions of eight arms, let alone the several hundred suckers, each itself mobile on a small, extensible stalk. Moreover, soft-bodied invertebrate animals do not have unusually large brains—rather the reverse. Even an octopus has a small central nervous system by vertebrate standards, although its brain is relatively large compared with other invertebrates. The implication from this lack of brain size is that such creatures—which include, for instance, all worms and mollusks—simply cannot compute the whereabouts of the ends of their flexible bodies in any considerable degree of detail. The very magnitude of the problem suggests that their motor control systems must, of necessity, be decentralized, with details of movements worked out by local reflexes, rather than in a central organization. In this event, the brain is required to order only the overall pattern of movements, and because of this it probably never receives detailed information about the position of the ends of its own body.

Similarly, the brains of many invertebrates never get precise information about the effort needed to carry out their orders. An octopus—as we have already seen—can lift a heavy object when the brain orders it to do so. But the sensory feedback that records the degree of muscle tension required apparently never penetrates to the highest centers, with the incidental result that the animal can never learn to detect weight differences.

A system that orders a response and assumes that the details will look after themselves may sound odd. Actually, even parts of human systems are so organized—our eyes are an example. If we decide to look to one side, our brain orders contraction of the appropriate eye muscles. The eyeball moves, and the image of the outside world passes across the retina. The world does not appear to move around us, because we know that we have moved our eyes, and we take this into account in assessing what we see. We know that we have moved our eyes, however, not because we can feel them move, but because the brain has, as it were, filed a copy of the order with the departments responsible for working out the significance of the retinal image. If eye movement is blocked—by attaching a sucker to a contact lens on the eyeball and holding it, for instance—this filed expectation is wrong. There is no movement of the image. But the brain has ordered a movement and expects the image to move, so a stationary image now implies movement; the world sweeps past. People subjected to this sort of experiment say that it is a most unpleasant sensation.

It seems probable that a great deal of the nervous system of soft-bodied animals is arranged in this manner—the activities of the various centers are integrated on a basis of copies of the orders to other parts of the nervous system, rather than on a sensory feedback from the motor machinery responsible for their execution.

If this is so, the thinking and learning part of the octopus's brain can be regarded as related to the neuromuscular machinery that deals with the animal's movement rather as I am to my car. I turn the wheel and press the pedals. I have no direct feedback from the front wheels to tell me that my order has been carried out, and I have no direct information about the events within the brake cylinders or the control of the gasoline mixture in the carburetor. I expect the details of the response to look after themselves.

Living machinery, fortunately, is a great deal more reliable than the machines we construct ourselves. It is not entirely unreasonable, therefore, for the thinking and learning part of the animal's brain to be organized on the assumption that its orders are always carried out, and for the animal to judge the effect of its orders from the consequent movement of the outside world, in much the same way as I judge the effect of turning the car wheel by the passing image of the external world that crosses my retinas. The octopus discrimination experiments yield results suggesting that this animal, at least, is organized with a brain that determines the broad outlines of movements to be carried out but has no part in, or knowledge of, organization of the necessary details.

The question is how to test the matter further. One way is to devise situations in which the animal's response to a sensory feedback will be different from the response to the afferent copy (a feedback in the central nervous system). The human eye example is one such test, and a related experiment can be done with the octopus—in this case to determine if the animal is aware of the position of its head and body as factors that will affect assessment of the retinal image.

For this experiment, the animal is first trained to discriminate by sight between two identical rectangles, one shown vertically, the other horizontally. It was found that the octopus, regardless of its position in the tank, continues to discriminate successfully between the rectangles. However, this apparently depends upon the eyes of the octopus remaining consistently oriented with response to gravity, so that the retinal image is itself constantly "right way up." Upset this "artificial horizon" and the animal ceases to distinguish between the rectangles. Experimentally, this can be achieved by removing the animal's balancing organs, the statocysts, which, among other things, produce the information that feeds reflex adjustment of eye position. After the operation, the slit pupil of the octopus no longer remains obstinately horizontal as the animal moves about; it varies, depending on how the animal happens to be sitting, and lies in the normal horizontal position only if the octopus sits squarely on the floor of its tank. When it sits on the side of its aquarium, the pupil (and therefore the retina) is at right angles to its normal orientation. Under these conditions an octopus that has been trained to take the vertical rectangle before the operation will reach out and grab the horizontal rectangle, which it had previously learned to leave alone. It correspondingly avoids the vertical rectangle.

It seems that no matter how long training is continued, the octopus never comes to realize that it always gets the wrong answer when it is sitting on the side of its tank. The implication, once again, is that sensory information about bodily position never feeds back to the parts of the octopus brain concerned in learning.

The most recent confirmation of the view that octopuses cannot learn about their own movements comes from experiments in which the animals must make detours to reach crabs that they can see on the far side of a partition in their aquariums. To get the crab, the octopus must go into the corridor, out of sight of its prey, and trek round into the feeding compartment, making an appropriate left or right choice at the far end of the corridor. There are various ways the octopus could manage this. It might, for example, simply creep along the face of the intervening wall, or it could (assuming for the moment that sensory information about its movements does reach the brain) remember the turn it had to make on going into the corridor. It is found, however, that an octopus does neither of these things. It solves the problem visually, by fixing the appropriate wall with its leading eye as it does into the corridor and hunting along for a break that will let it turn in the direction of the crab. We know this from the behavior of octopuses that have been blinded in one eye. After the operation, which in no wise interferes with their normal locomotion or their interest in crabs, the animals continue to detour successfully in one direction, but fail when required to collect crabs seen on the other side. They fail, it seems, because they go into the corridor and fix on the wrong wall; working along this one leads them into the wrong feeding compartment. They seem unable to learn to correct for this type of error.

The results with animals blinded in one eye are interesting because the errors are by no means inevitable. The octopus normally uses one eye at a time, anyway, and individuals blinded on one side do succeed from time to time in making perfectly satisfactory detours. One might expect them to learn by experience that getting food follows their making only a turn toward the wall when they go into the corridor. But instead they continue to make repeated unsuccessful runs 180 degrees off course. On these occasions the octopuses detour quite as rapidly—one is tempted to say "confidently"—as usual, apparently unaware that an essential movement sequence is missing from their response. The implication, again, is that this animal is unable to take its own movements into account when it learns to carry out various activities.

The detour experiments, in short, confirm the thesis arising from the tests on visual orientation and tactile discrimination. For these soft-bodied animals, learning on a basis of clues from sense organs detecting external events is one thing, and possible; learning from internal receptors is quite another, and apparently not.

This conclusion has a variety of consequences for the study of animal behavior. If true, it means, for a start, that there can be no question of animals such as octopuses learning to make skilled movements, since they cannot learn to make progressive small modifications in the way they manipulate. All the motor machinery is there, but the sensory inputs that would be needed fail to penetrate to the upper levels of the central nervous system. Operant conditioning, generally, presents difficulties. The means by which soft-bodied animals find their ways about become interesting because we must discard the possibility that the animal has anything in the nature of a map sense, based on the direction and distance between objects in its environment. If an octopus (or a worm, or a limpet) has no memory of the detail of movements that it has made or been obliged to make on the outward journey, it cannot possibly compute its way "home" by any form of dead reckoning.

We are all used to the idea that animals may differ in their capacity to learn. It would also appear obvious that the range of things that they can learn is limited by the properties of their sense organs. It is perhaps a less familiar idea to suggest that their performance may be limited, not so much by their sensory instrumentation as by the way their central nervous systems are organized to deal with the information collected. The octopus is an animal in which the point is clear precisely because the animal learns to solve so many problems so quickly; its failures stand out against a wide background of things that the animal can be taught to do. The sensory instrumentation is there. The muscular machinery to carry out a variety of tasks is present. The animal's performance can be attributed only to the organizational consequences of flexibility.

There is no reason to believe that the octopus is exceptional in this respect. It seems safe (or if not quite safe, at least interesting) to predict that with increasing knowledge of the behavior of animals, two broad categories will be revealed. There will be those animals, including humans and perhaps the arthropods, that can learn to modify their behavior by taking into account details of the movements they make. And there are the rest—all the soft-bodied animals, among which is the octopus—that cannot.

At the present time we know more about the behavior of arthropods than all the other invertebrate animals put together. One reason is that arthropods can be made to perform well in training experiments—maze tests and so on—that seem simple to a vertebrate. The other invertebrate animals often seem curiously dim by comparison, and in most cases attention quickly returns to the apparently more promising band of creatures with joints. Work on the octopus has shown, however, that soft-bodied, flexible invertebrates may be surprisingly quick to learn, provided only that the experiments are appropriate to their particular types of organization. One can perhaps hope for equally interesting results elsewhere among the invertebrates, as more sophisticated methods of testing their performances are devised. It is the purpose of this article to call attention to one of the complexity of factors that must be taken into account in such studies.

PART 2
EVOLUTION AND BEHAVIOR

Most of us have experienced the unpleasant consequences of motion sickness, precipitated by travel in boats, airplanes, or even automobiles. This powerful vestibulogastric illness, characterized by nausea, vomiting, pallor, and sweating, is disabling, unpleasant, and all too common. And judging by its presence in such diverse animal groups as monkeys, horses, sheep, some birds, and even codfish, motion sickness is undoubtedly a bodily reaction that arose quite early in vertebrate evolution. But a bodily reaction to what? In homeostatic processes such as pollen allergies, a foreign antigen (e.g., ragweed) elicits violent sneezing which helps to expel the noxious stimulus. But motion sickness has no such adaptive function. Vomiting doesn't remove the offending stimulus, and doesn't alleviate the symptoms. Why then did it evolve in the first place? The answer, surprisingly, may have more to do with eating than with traveling.

Most species of animals possess mechanisms for protection against poisonous foods. For example, some populations of monarch butterflies contain cardiac glycosides that are extremely toxic. When hungry, naïve bluejays are fed monarch larvae, the birds become ill and vomit within 30 minutes after ingesting the poison. Furthermore, a single encounter with a toxic insect larva is frequently sufficient to produce permanent avoidance behavior by the jays. In more recent studies, when comparable toxins were fed to human subjects, they also vomited, perspired profusely, and reported sensations of dizziness. So the hypothesis is that motion sickness is an adaptive response evoked by an inappropriate stimulus. It evolved as a mechanism for expelling noxious substances from the body (vomiting) and for conditioning the individual to avoid that substance in the future (sweating and dizziness). When you are traveling on a ship, the rocking motion causes the vestibular system to produce a sensation of dizziness. The brain incorrectly interprets this information by concluding that you once again have been dining on those noxious insects (or other poisonous foods). The emetic reflex is triggered, and you promptly throw up!

The above (admittedly unpleasant) introduction to this part of the book illustrates that evolutionary hypotheses about the origins of behavior can be generated even when no fossil record is available for corroboration. Unlike com-parative morphologists, animal behaviorists must rely principally on evidence derived from analyses of existing species. Although extant species obviously do not represent an evolutionary sequence, particular behavioral characteristics are often "frozen" at a stage that represents an evolutionarily earlier pattern. The articles by Collias on bird nests, by Kaston on spider webs, and by Kluger on fever all illustrate this comparative approach to unraveling the phylogeny of behavior.

While it is generally recognized that behavioral patterns within a lineage, like structures, can change drastically under the influence of natural selection, we must also remember that evolution is two-sided. Behavior not only evolves, but actually can influence the future course of evolutionary change. Mating in the animal world is rarely random. Individuals often choose mates on the basis of conspicuous structures (e.g., colors of bird plumage) and elaborate courtship rituals. Above all, these adaptations serve as reproductive isolating mechanisms, ensuring that mating occurs only among members sharing a common gene pool. In the article by Cooke, we are provided with a first-rate example of how color polymorphism in snow geese influences behavioral reproductive isolation. Finally, Martin compares re-productive physiology with parental behavior, which he sees as a key to elucidating primate evolutionary history.

8

Evolution of Nest Building

NICHOLAS E. COLLIAS

The lives of birds usually center about the nest during the most active part of their existence. The importance of the study of nests was recognized by ornithologists of a past generation in the creation of a special term, "caliology" (from Greek *kalia,* hut or nest, + ology), for this area of scientific investigation. As a rule, however, nest collections of most museums compare very poorly indeed with the collections of bird skins to be found in the same institutions.

With the growing emphasis on the study of the living bird, there is a renewed emphasis both on the study of nests and on the behavior patterns associated with their construction and use. There is also an increasing recognition that nests and nest sites often bring to a focus the principal habitat requirements of a species. The extreme variations in nest form, structure, and elaborateness possess a significance that, to be understood, often requires close study of the bird and its nest under natural conditions. Furthermore, like other characteristics of species, the type of nest built depends on an evolutionary history to which we may gain some clue by means of comparative study of related species. A nest may be defined as an external structure that contains eggs and young and that aids in their survival and growth.

Building a nest often requires a great deal of time and energy in many species of birds. It is common for them to make a thousand trips or more to gather and transport all the necessary materials. Natural selection may be expected to favor any behavior patterns that economize on undue effort, provided that some crucial advantage of the species is not thereby sacrificed. For example, theft of nest materials from other members of the species is very common among birds, but it is obvious that undue use of this method by all the individuals of a species could have a serious effect on the reproductive rate at the population level. For example, years ago Alexander Skutch estimated that up to half the nesting failures he observed in a population of Rieffer's Hummingbird (*Amazilia tzacatl*) in Central America resulted from the theft of materials from nests of other hummingbirds, resulting in their collapse, with eggs or nestlings. It is evident that the necessity to protect the nest itself has been an important force in the evolution of what is commonly called territorial behavior.

Competition between closely related species often results in the evolution of great differences in habitats and nest sites—differences that may give the birds their names. Compare, for instance, Barn, Cliff, Cave, Tree, and Bank Swallows, all well-known species in North America. In turn, differences in the nature of the substrate for the nest imposes special engineering requirements with regard to materials, form, structure, and placement of the nest for each species.

The primary and general functions of a bird's nest are to help ensure warmth and safety for the developing eggs and young, but if the evolutionary forces involved in nest-building characteristics of any avian species are to be fully clarified, one must often be familiar with other aspects of its life history as well. Convergence in nest form and structure between unrelated species often furnishes significant clues to the nature of the ecological forces operating in the evolution of the characteristic nest.

Our main emphasis is not to use nest characters to help develop particular phylogenies in groups of related species, but rather to attempt to show how one can analyze the ecological nature of the selection pressures that have led to the evolution of the main types of nests. Of course, the problems of warmth and safety for the young are generally most acute for small birds and their young, which explains why, as a rule, birds of small body size build nests that are more elaborate and better concealed than are those of larger birds.

When birds evolved from reptiles they probably developed the ability to maintain a high, constant body temperature more or less coincidental with the ability to fly. It seems likely that during the transitional evolutionary period, when flying ability and temperature regulatory mechanisms were being perfected, many birds became torpid during very cool nights. A few birds known today, such as the Poor-will (*Phalaenoptilus nuttallii*), still do so. The probable imperfection of body temperature control in ancestral birds is an argument in favor of the theory that the first birds did not incubate their eggs by sitting on them, as do most modern birds, but probably adopted some other means. Perhaps they buried their eggs in the soil and relied on heat furnished either by decaying vegetation or the sun. Many reptiles do this, as do birds of the family Megapodiidae. Within the confines of one genus, *Megapodius*, the nest may vary from a simple, small pit dug in the sand, large enough for just one egg, to gigantic mounds of soil and decaying vegetation from 30 to 60 feet long and reaching 15 feet in height—the largest bird nests known.

Some of the megapodes have developed an efficient control over the temperature in their nest mounds far beyond that seen in any reptiles. One megapode, the Mallee Fowl *(Leipoa ocellata)*, lives in arid regions of Australia where temperatures vary from below freezing to above 38°C. (100.4°F.), and where even in midsummer the nighttime temperature may be 17°C (30.6°F) cooler than the day temperature. Yet, by varying the depth of the soil over the eggs and thus the degree of insulation from cold or of exposure to the sun, this bird maintains a relatively constant incubation temperature of between 32°C. (89.6°F.) and 35°C. (95°F.), as H. J. Frith discovered by making careful measurements throughout the breeding season.

At the close of the Mesozoic, the climate changed from humid tropical or subtropical to drier conditions, with greater extremes in temperature. This was probably met in early avian evolution by two different solutions to egg incubation. Some birds were or became mound builders, and evolved considerable efficiency in regulating the temperature in the mound about the eggs. Other birds had or developed the method of incubation by application of parental body heat to the egg.

Once birds had evolved the ability to maintain a high body temperature throughout the night, a strong selection pressure for direct parental incubation of the eggs would be established coincidentally. Eggs of reptiles often take months to hatch, whereas those of birds frequently hatch within a matter of weeks. Total predation on the eggs would diminish as the developmental period was shortened. The danger of predation from various nocturnal enemies, especially from the small contemporary mammals, would favor the habit of staying with the eggs and defending them, if necessary, during the night. F. H. Herrick has suggested that the origin of incubation by sitting on the eggs probably arose from the tendency of birds to conceal them as a protection from potential predators.

Use of natural or excavated cavities is common among birds and, in some instances, has evolved into quite elaborate excavations in the ground, in banks, or in trees. Almost half the orders of birds recognized by Ernst Mayr and Dean Amadon in their classification of the birds of the world contain some species that nest in cavities. Whole orders of cavity nesters are represented by the kiwis, parrots, trogons, coraciiform (kingfishers and their relatives), and piciform (woodpeckers and their relatives) birds. The habit of nesting in cavities furnishes considerable shelter and safety, particularly to small birds. Populations of House Wrens *(Troglodytes aëdon)*, for instance, have often been increased when a good supply of nest boxes has been put out for them. The shelter and safety of cavities has resulted in intense competition for these nesting sites. The House Wren discourages competition by other birds such as the Prothonotary Warbler *(Protonotaria citrea)* by puncturing the warbler's eggs. Along with each size class of woodpecker goes a host of other species that compete with them for the corresponding size of nest cavity. The European Starling is notorious in this regard. I have seen a starling in Ohio seize a Yellow-shafted Flicker by the tail and cast it out of the flicker's freshly dug tree hole in which a pair of starlings subsequently reared a brood. The German nature photographer Heinz Sielmann observed that when a European Nuthatch *(Sitta europaea)* takes over a tree cavity it forestalls its chief rivals, the starlings, by collecting mud from nearby puddles and plastering it around the entrance to the tree hole until the entrance is so small and narrow that, while the nuthatch can slip through, the starling cannot.

Different stages in the evolution of nest sites in tree holes are represented by different species. Some use natural cavities, others modify them or excavate cavities in soft or decaying wood or even in hard, living trees, as do some of the larger woodpeckers. Similarly, in the case of birds that nest in holes in the ground, various degrees of specialization are illustrated by different species. These range from a shallow scrape in many ground nesters, to a relatively short burrow like that of the Rough-winged Swallow *(Stelgidopteryx ruficollis),* to larger burrows such as those made by the related Bank Swallow *(Riparia riparia),* which may be six or more feet long, tunneled in something of an upward course and providing protection against driving rain.

The evolutionary climax of excavated nests is the construction of nesting cavities by certain birds inside the nests of social insects. V. A. Hindwood has listed 49 species of birds, including kingfishers, parrots, trogons, puffbirds, jacamara, and a cotinga, that are known to breed in termite nests. In fact, some 25 percent of the species of kingfishers of the world nest in termitaria. As the excavation by the bird progresses, the termites seal the exposed portions of their nest so there is no actual contact between birds and insects. Many birds that breed in termite nests do not normally eat the insects of the colony in which they are nesting. Birds that breed in nests of social insects are all from taxonomic groups characterized by nesting in cavities, and cavities in old and deserted termite nests are at times used by birds that normally breed in earth banks or tree holes, suggesting how the habit might have evolved.

Nesting in a hole goes a long way toward meeting the essential nest functions of warmth and safety, and thereby actually tends to block further evolution of truly elaborate increment nests built up from specific materials. In fact, nests built within cavities may undergo a regressive evolution. One can see all degrees of increasing simplification and reduction from an elaborate roofed nest to a mere pad, as in the case of the Old World Sparrows (Passerinae) that nest in tree holes.

In contrast to cavity nesters, birds that build open nests on the ground are subject to a greater chance of nest failure. Consequently, there is a strong selection pressure to build an adequate nest, to develop other special means of parental care, or to evolve markedly efficiently concealing coloration. Although with the origin of direct parental incubation it was no longer necessary to dig a pit for the eggs, most birds that nest on the surface of the ground today still begin by making a circular scrape with their feet while crouching low and turning in different directions. This hollow may then be lined with various materials to protect the eggs from the cold, damp ground, while a rim of materials around the body of the sitting parent provides added insulation for the eggs. The materials are pushed to the periphery of this rim and built up into a circular form by much the same sort of movements of the feet and body as are involved in making the initial scrape in the ground. Many ground-nesting birds prevent the flattening down of the peripheral raised rim of the nest by repeatedly reaching out with the bill and drawing materials in to the breast or passing them back along one side of the body before dropping them. These patterns of making a nest can be seen, for example, in the Canada Goose.

Nests built on the surface of the ground are especially liable to be flooded. These nests are often built on slight elevations and, as in the case of the Canada Goose, may be built up higher during a flood. The Painted Snipe *(Rostratula benghalensis)* in Australia may lay its eggs on the bare ground when the earth is dry, but if it is covered with water, a solid nest of rushes and herbage is made. Similarly, the Adélie Penguin *(Pygoscelis adeliae)* builds up its nest of small stones if thaws cause flooding. In the Antarctic, where this penguin nests, William Sladen noticed one nest that had a stream of ice-cold water running through it. The male on the nest kept reaching forward, collecting and arranging stones about himself and his half-submerged eggs. By the next day the nest and eggs were above water, and eventually the eggs hatched.

Parental behavior may supplement or even substitute entirely for a nest under severe environmental conditions. In the Arctic, persistent close incubation by the parent bird is necessary and characteristic, regardless of whether or not the nest is well insulated. It has been found that the Semipalmated Sandpiper *(Ereunetes pusillus),* which builds no nest, keeps its eggs as warm as do other Arctic species that have substantial nests. The Emperor Penguin *(Aptenodytes forsteri),* which breeds in the Antarctic winter, has no nest, but rests its single egg on its feet, covers the egg with a fold of skin from its abdomen, and incubates it against the body. Probably no other animal breeds under such trying conditions. At an opposite extreme, eggs or nestlings exposed to strong tropical or subtropical sun in open situations are customarily shaded by the body and wings of the parent, as in the case of the Sooty Tern *(Sterna fuscata)* of Midway Island, whose nest is a mere scrape in the coral sand.

Birds nesting on the ground are subject to high predation, and it is no accident that the classical cases of concealing coloration are found in birds, like the ptarmigan, which lay their eggs in an open nest on the ground. In certain cases, the color pattern of the eggs and the young, as in the European Stone Curlew *(Burhinus oedicnemus)*, or of the young and the parent, as in the Whip-poor-will *(Antrostomus vociferus)*, matches the surroundings so closely that the nest has disappeared in evolution, presumably because a nest itself would attract attention and be too conspicuous.

The dangers of ground nesting and the intense competition for tree holes have apparently provided a strong selection pressure for the evolution of increment nests placed on the branches of trees or, in some cases, against the faces of cliffs. Species of birds with prococial young (covered with down and able to move about) are generally ground nesters, whereas species with altricial (naked and helpless) young frequently nest in trees or bushes. In the prairie country of northwestern Oklahoma there are few trees, and R. L. Downing found that those Mourning Doves *(Zenaidura macroura)* which nested in trees were almost twice as successful in fledgling young as were those which nested on the ground. There was a definite preference among the doves here for nesting in trees. In fact, Margaret Nice, also of Oklahoma, found long ago that pairs of Mourning Doves nesting within forks of trees had a greater success than did those pairs nesting farther out on branches. Conversely, under safe nesting conditions, certain species of birds (such as the Osprey and Robin on Fardiner's Island, New York) that normally nest in trees may nest on the ground, thereby conserving the energy required to fly up into a tree with nest materials or with food for nestlings.

Tree nesting requires the solution of new types of engineering problems. The nature of the materials used varies with the body size of the bird and its lifting power. Large birds use large twigs and even branches, which will not readily be blown out of trees by the wind. Medium-sized birds use small twigs or grasses or both, sometimes adding mud to help attach and bind the materials. Many small birds use spider or insect silk as a binding material for the attachment of the nest to the substrate and to bind various other materials of the nest together.

The platform nests of large birds, such as the American Bald Eagle *(Haliaeetus leucocephalus)* and the European White Stork *(Ciconia ciconia)*, may have twigs and branches added year after year and may become very large and very old. F. H. Herrick describes an eagle nest 12 feet tall and 8½ feet across, and estimates that it weighed over two tons when, in its 36th year, it fell during a storm, together with the tree. F. Haverschmidt has managed to date back to 1549 one White Stork nest that was still in use in 1930.

It seems probable that every type of material characteristic of the nest of a given species has a definite function, and that the proportions of different types of materials used vary not only with availability but also with the requirements of particular substrate and habitat conditions. Otto Horvath observed that Robin nests in British Columbia contained more mud when the birds had to use short building materials, more tough and flexible rootlets when the nest was in an especially windy spot, and more moss when it was in a relatively cold microclimate.

Cup nests of very small birds are likely to be heavily insulated, as is true of the nests of most species of hummingbirds. There is some evidence that, compared with lowland species, hummingbirds that nest in high mountains build nests with relatively thick walls or seek the protection of caves.

Nests attached to the vertical faces of cliffs, caves, or buildings furnish protection against nonavian predators, but pose special problems for attachment of the nest. The swifts have generally used adhesive saliva, while the swallows have evolved toward more frequent use of mud, probably with some admixture of saliva. Different species of swiftlets (*Collocalia*) can be arranged in a graded series from such species as *Collocalia francica*, which make nests using pure saliva (the source of the ideal bird's-nest soup of the Chinese), through other species that use various admixtures of plant materials, to those that build nests of more conventional types. The nest cement of *Collocalia fuciphaga* is sparse and soft, and the nest, which is composed largely of moss and other plant materials, can be placed only on an irregularity in the cave wall that will take all or a good part of the weight of the nest. In contrast, nests of other cave swiftlets can be glued to vertical walls in the cave.

Building of a roofed increment nest is very rare among nonpasserine birds, whereas almost half of 82 families and distinctive subfamilies of passerine birds recognized by Mayr and Amadon construct roofed nests or contain species that do so. Although roofed nests, aside from use of natural cavities, are unusual among passerine birds of the North Temperate Zone, they are typical of many tropical genera and families of passerines. In most instances, birds having nonpensile roofed nests first make a basal platform and then build up the sides and roof, and this sequence of building suggests that roofed nests probably evolved from nests that were at one time open above.

Roofed nests may be made of very different materials by different birds—woven or thatched of grasses in many weaverbirds, of plant fibers in certain icterids, of short, heterogenous plant materials bound by spider silk in sunbirds and some titmice, and of mud in Cliff Swallows. The convergent evolution in such diverse instances is evidence of the great importance of a roof in the life of small nesting birds.

The roof of domed nests is important in shading young birds from the sun. Solar radiation is most intense in the tropics and would quickly kill small, naked altricial nestlings exposed to direct rays. The Galapagos Finches (Geospizinae), for instance, have an equatorial habitat and, unlike most Fringillidae, build roofed nests.

One function of the roof of domed nests must be to shed rain. Most small birds in the tropics nest during the rainy season when insect food is abundant. Skutch observed that the nests of the Yellow-rumped Cacique *(Cacicus cela)* in Central America are all open at the top during the early part of the breeding season before the rains. But as the rains begin, after the eggs have been laid, the top of the entrance is gradually roofed over and the nest entrance becomes a bent tube opening downward.

Protection from bird and mammal predators is aided by placement of nests in dense cover, especially in thorn trees. The buffalo weavers of Africa build a thorny cover or shell to the nest. Furthermore, the Whiteheaded Buffalo Weaver *(Dinemellia)* is famous for placing thorny twigs along the boughs leading to its nest.

Among the predators of nestlings, snakes are more numerous and varied in the tropics than in colder localities, and perhaps roofed nests help deter snakes as well as other enemies. The weaverbirds all build domed nests; in addition, the nests placed in trees tend to evolve a firm pensile attachment and a bottom entrance with a long entrance tube, enhancing protection from snakes. In East Africa, Van Someren once watched a green treesnake trying to get at the young in a Spectacled Weaver's nest. The snake negotiated the slender, pendent branch and reached the nest, but could not manage the 12-inch tubular entrance and fell into the pond below.

A woven construction facilitates evolution of roofed and pendulous nests and enhances the nest's coherence. A whole series of representative stages can be traced in the weaverbirds from loose, crude, irregular weaving to the close, neat, and regular pattern to be found especially in those species that build pendulous nests with long entrance tubes. The nest of Cassin's Malimbe *(Malimbus cassini)*, a black and red forest weaver of Central Africa, is perhaps the most skillfully constructed nest made by any bird.

In contrast to its importance among the social insects, a compound nest has been evolved by only a few species of birds. The compound nest consists of a common nest mass in which more than one pair of birds or more than one female of the same species occupy separate compartments. Such birds include the Palm Chat *(Dulus dominicus)* of Haiti, the Monk Parakeet *(Myiopsitta monacha)* of Argentina, the Black Buffalo Weaver *(Bubalornis)* of Africa, and the Sociable Weaver *(Philetairus socius)* of the Kalahari Desert in South Africa.

The nest masses of the Sociable Weaver have often been compared to haystacks in a tree. These nests are not woven, but are thatched with dry grass stems. Each nest mass is often several feet thick, of irregular extent, and may be over 15 feet long in the longest dimension. The top of each nest mass is dome-shaped, the underside relatively flat and riddled with scores of separate nest chambers. Many different individuals may work together on the common roof, which may be one key to the evolution of this remarkable structure. The roof enhances protection from predation for all, as does the outer thorny shell in nests of the Black Buffalo Weaver or the projecting eaves in those of the Monk Parakeet.

Special security from predation seems to be an important factor in making possible gregarious breeding, a phenomenon that is rather rare among small land birds, although common in sea birds on remote islets or inaccessible cliffs. One well-known example of gregarious nesting among passerine birds in an obviously safe nest site is that of the Cliff Swallows, whose mud nests are placed on the vertical face of a cliff or building. Colonies of the Sociable Weaver are frequently found, in camel's thorn acacia trees, in which each of the many thorns may contain a colony of ants *(Natural History,* January, 1965). The habit of nesting in association with colonies of noxious insects is found in many African birds, and may have been a predisposing force in the evolution of the Sociable Weaver's compound nest. Possible intermediate stages are represented by the nests of a related species in East Africa, the Grey-headed Social Weaver *(Pseudonigrita arnaudi).* This gregarious species breeds in small colonies in which the different nests may be well separated, but when placed in ant-gall acacias, whose thorns give added security, many of the nests are grouped into common masses. We have counted up to nine nests in one mass.

When one considers the extent of variation in birds' nests there seems a vast difference between, say, the shallow scrape of a Sooty Tern in the coral sand of a tropical island and the immense communal dwellings of the little Sociable Weaver or the exquisitely woven cradle of Cassin's Malimbe. These variations are an exciting challenge to the ornithologist who would attempt to explain their origin in terms of the complex ecological and behavioral forces that have shaped the nests of birds in the course of evolution.

9

Evolution of the Web

B. J. KASTON

When you have seen a spider web, or brushed against one accidentally, have you ever wondered how the phenomenon of silk use in spiders developed? Of course, we do not know the entire story, but from the studies carried on by many observers, some parts of the tale can be seen to fit into a pattern. For example, some spiderlings build snares that differ from those they construct as adults, and because the webs of these youngsters are less specialized, they more clearly show resemblances to their ancestors' webs. Here is one clue to our reconstruction of the story of web evolution.

Although only some spiders are web builders, all produce silk. Chemically, silk is a polymerized scleroprotein, and that of spiders varies somewhat in its physical properties according to which of several glands produce it. These glands all lie in the abdomen, and the silk is emitted through ducts that discharge onto small appendages known as spinnerets, located at or near the posterior end of the abdomen, just in front of the anus. Most spiders have three pairs of spinnerets, provided with many small spinning spools and spigots of several kinds and sizes; the silk issues through their openings.

Spider silk is of two general types—sticky and nonsticky. The non-sticky is presumably the more primitive; it is made by all spiders and forms the structural foundations of the webs. Sticky silk is of two types, depending upon whether the maker belongs to the Ecribellates or Cribellates—the two major groups of higher spiders. In the Ecribellates, the viscid silk appears as sticky globules along the fiber, but in the Cribellates it is flocculent.

Cribellate spiders have special silk glands, with ducts opening through numerous fine pores onto the cribellum, a sievelike plate lying just in front of the spinnerets. The cribellate spiders also have, along the dorsal surface of the metatarsal (the penultimate) segment of the hind legs, a single or double row of curved bristles, which compose the calamistrum. This is used to comb out the silk that issues from the cribellum, thus forming the flocculent threads. Ecribellate spiders lack the calamistrum (as well as the cribellum, of course), and their viscid silk is deposited as microscopic globules on dry threads.

Two theories have been proposed to account for the evolution of silk. Since spinning arachnids other than spiders (the whip scorpions, false scorpions, and mites) have silk glands emptying at the mouth region, it has been suggested that spiders originally discharged their silk in similar fashion. The "spitting spider," *Scytodes,* still retains this trait, flinging a gummy material from its chelicerae. This is in addition to the more usual spinning apparatus used in producing the egg sac silk. Besides ordinary silk, the trap-door spiders use material from their mouths to cement particles of earth together to build their burrows.

According to a second point of view, silk was originally excretory material deposited behind as the spider ran about, and in some way this became the characteristic drag line that is trailed after the spider. When one considers the close proximity of the spinnerets to the anus and the way in which some, like the ground spider *Zelotes,* use excrement, in addition to silk, to fashion egg sacs, this theory seems plausible enough.

Spiders use their silk in many ways other than for web building. One such use—for covering eggs—suggests something about the evolution of the web. All spiders, as far as is known, cover their eggs with silk, and because mites, whip scorpions, and false scorpions also cover their eggs in this manner, one line of thought suggests that spiders first used silk to produce an egg sac, and that the web itself later evolved from a mass of threads distributed around the sac. While the spider hid with its egg sac in a crevice or hole, the silken drag line, which was emitted as she moved about, formed a lining for the shelter, and this eventually became tubular in shape.

Some spiders still build only this little tube, in which they hide, while others, in addition, have constructed one or more flaps, or doors, from which they can emerge to seize insects that touch or pass the flaps closely. There is a very primitive spider, *Liphistius,* that provides us with evolutionary evidence of other spiders. Its tubular, subterranean snare, from which we can trace the cribellate and ecribellate webs, shows us how the next step has been reached. In addition to a thin flap, or trap door, at the top, and a smaller flap at the opposite end to serve as an emergency exit, seven long threads extend from the rim of the upper aperture. Together these resemble radial threads of an orb web.

The further development of webs may be shown with *Liphistius* at the base and Cribellates and Ecribellates radiating from it (figure 8.1). Web evolution in these two groups has progressed along similar, but independent, lines—a phenomenon often referred to as convergence.

In general, the more primitive spiders live in, on, or near the ground, while the most highly developed web, the cartwheel-shaped orb, is aerial. Thus it is supposed that in the history of web building, ground types appeared first and were later followed by those constructed in space. The first webs made by spiders were of nonviscid silk, and the spider became aware of the presence of an insect when it hit the web and produced a vibration. When viscid silk was later added, the web also functioned in trapping the prey.

The sequence in which an orb web is spun has often furnished the observer with information that has been used in setting up the still-hypothetical tree of web evolution. The account that follows is based upon what is done by a mature female, since adult males often spin no webs and may even live out their relatively short lives on the web of a female.

The spider first ties together the objects between which the web is to be spun. This nearly always involves use of air currents to carry the thread. The spider tests the line emitted into the wind to determine when it has stuck. It then pulls this bridge line taut and fastens it. The spider crawls across the line, paying out a heavier thread that will serve as the bridge line. It then spins a frame consisting of foundation lines, which, together with the bridge thread, determine the plane of the web. The next step is to put in the radii, and in order that they may sustain equally the stretching of the web, they are connected by means of a few complete turns, or laps, of a spiral thread spun just outside the hub. This area is called the attachment zone. Once the hub and attachment zone are completed, the spider spins a spiral thread, starting at the end of the attachment zone spiral and continuing to the periphery, with turns as far apart as the animal can stretch. The scaffold spiral is temporary, and its function is to hold the radii in place during subsequent operations.

Now is begun the construction of the actual snaring spiral of viscid thread. This is always spun from periphery to center, but stops some distance short of the attachment zone, leaving an area called the free zone. The original scaffold spiral is cut away, turn by turn, as the viscid spiral approaches it. Once the catching spiral is finished, the spider has only to make a change in the hub. Some bite out the threads from the center, so that the hub becomes an open one. Others cover the hub with a dense sheet of silk. Some insert a band of silk above, or both above and below, the center. This "decoration," usually called the stabilimentum, may be straight, zigzag, or, in some cases, circular.

Although we began the account of the web's evolution with *Liphistius,* we must remember that the early arachnid from which spiders arose probably possessed no silk apparatus at all, but depended upon speed and agility in hunting and capturing its prey. Presumably, there then appeared those that used silk only to cover their eggs, and representatives of this type can be found today. Some mothers even carry the newly hatched spiderlings on their backs as they run about. But there were those that hid with their egg sacs in a crevice, which eventually they lined with silk. This was followed by the elaboration of a door with fibers radiating from it. This entire bit of history can be seen in one group—the lycosids—which will be discussed in conjunction with the Ecribellates.

Let us first consider the cribellate line. We must conjecture that the earliest of these probably spun only some sort of tube. In the European *Eresus,* this tube is partly underground and partly over the surface, with a small sheet extending to one side. A larger sheet is constructed by the members of the family Psechridae. In the widely distributed *Amaurobius,* the tube is poorly defined, and there is a mass of threads extending in all directions from the lip. In *Filistata,* of warm climates, the tube is better defined, and the arrangement of threads extending from its mouth suggests the regular placement of radii in an orb. The situation is somewhat similar to that seen in the web of *Liphistius,* but the door to the tube is lacking.

In addition to the tubular portion, the Old World *Stegodyphus* has lines of silk arranged to resemble a sector of an orb, and these lines are irregularly banded with cribellar frets. *Sybota,* of Europe, proceeds further: the tube has disappeared, and the web, although no longer on the ground, is not completely in space; the hub of the orb is always attached to a supporting twig or stalk. This would seem to be an intermediate condition between *Filistata,* whose web is entirely against a solid surface, and *Uloborus,* whose web is out in space. Another peculiarity of the web in *Sybota* is that no fold spiral remains, and the cribellar silk is fastened to frame threads and radii. *Uloborus* produces a full orb, with the sticky cribellar silk on the catching spiral, and with a stabilimentum. Yet only a scaffold spiral is found in the web it spins as a young spider, thus indicating a link with *Sybota.* Here, nicely displayed, is one of the evolutionary clues.

The phenomenon of reduction is again encountered in the webs of *Hyptiotes* and *Miagrammopes.* We recall that in *Stegodyphus* the web included a tubular portion and a portion looking much like a sector of an orb. If we lift this sector away from the tube and suspend it in space between twigs, we will have something resembling the web of *Hyptiotes,* the common, triangle-web spider of the woodlands. The web of the latter consists only of a triangular sector composed of four radii, which converge at the apex to form a single thread that is fastened to a nearby twig. To capture her prey, the spider, having fastened herself to the twig with a drag line, holds taut the single thread attached to the four radii and releases the tension with a spring when an insect strikes the web. In the tropical stick-spider, *Miagrammopes,* all that remains of *Hyptiotes'* sector is a single thread about a meter long. For somewhat less than half its length, the central part is covered with cribellar silk. This single line likewise functions as a spring trap in exactly the same way as that of its presumed evolutionary predecessor, *Hyptiotes.* The spider releases the tension when some insect alights on the thread and thus entangles the prey.

The last group of Cribellates to be discussed belongs to the family Dinopidae. These uncommon spiders construct a rectangle or trapezoid of dry silk, across which are spun a number of bands of cribellar silk. The spider holds the resulting snare, 10–20 by 15–25 mm. (or slightly larger), with its front legs. When an insect approaches, the spider moves its front legs apart, thus expanding the elastic snare to between 5 and 10 times its original size, and may even hurl the net over its victim. One can only conjecture as to the origin of this device.

Since the Ecriballates are more numerous and variable than the Cribellates, the account of web evolution in this group will be more lengthy. Here, too, we go back to *Liphistius* with its burrow. Many other spiders live all their lives in such burrows, with or without door flaps, and many either live this way while carrying their eggs, or place the eggs in such silk-lined holes. Some go a step further, extending the seven radiating threads into a sheet around the mouth of the burrow.

Present-day members of the wolf spider family, Lycosidae, show all the stages. For example, *Pardosa* is a hunter, and even at the time of impending motherhood roams the field stalking her prey with her egg sac attached to her spinnerets. Upon emergence from the sac, her children make their home on her back, where one may see them clustered like a bunch of grapes. Several of the *Lycosa* species make use of a silk-lined burrow, but only at the time of egg laying. Species of *Geolycosa,* on the other hand, spend all their lives in such burrows. It has been reported that some wolf spiders, such as *Lycosa aspersa,* even build doors for the burrows, while *Pirata* builds a tubular nest among moss plants over water. Finally, *Hippasa* spins a sheet of silk around the entrance of the burrow.

The best-known of the tunnel builders that fashion doors are the trap-door spiders, members of the family Ctenizidae. Some of these make thin "wafer" doors at the top (again like *Liphistius*), often with another such door at the entrance to a side chamber below. Later, the doors were strengthened so they were thick and fitted like a cork; they usually are held shut with surprising strength.

Another line of descent led to the typical tarantulas (Theraphosidae), many of which have temporary or permanent silk-lined, shallow burrows in the ground. A few have become arboreal, while members of the closely related, tropical Dipluridae construct sheet webs over the ground, with one side fashioned into a funnel under a rock or in a crevice.

Paralleling the evolution of the theraphosids, ctenizids, and diplurids is that of the group of so-called atypical tarantulas. These display an even closer structural relationship to *Liphistius*. Some, such as *Hexura*, of our Pacific Coast states, run over the surface of their sheet webs. Others include the "folding door tarantulas," whose burrows are closed with a door composed of semicircular halves. And in this group we have the purse web spider, *Atypus*, whose web is a combination of a subterranean and a surface portion. First, the spider constructs a small horizontal cell—much like that of a jumping spider—on top of the ground, and from this it extends the tube downward. It has been suggested that the whole merely represents an elaboration of the ancestral cocoon.

Returning to *Liphistius*, we recall that a number of threads radiate from the lip of the burrow. Some of the more primitive spiders construct, in a crevice of rock aboveground, a silken tube with a similar arrangement of threads radiating from the mouth, although there is no door. Gradually, by the addition of many more radial threads, a kind of funnel was perfected around the entrance of the tube. Thus we have the kind of snare found in *Coelotes* today. The extension of the lower side of this funnel into a sheet gives us the typical agelenid web.

An added refinement to the funnel webs of many agelenids is the network of irregular threads above the sheet. This "stopping maze" catches flying insects, causing them to fall on the sheet, where the spider seizes them.

Let us now consider the ancestral spider that hunted on shrubs and herbage, and hung its cocoon among vegetation while remaining on guard. It seems reasonable to suppose that haphazard strands crossed one another at all angles, but radiated outward from the vicinity of the egg sac. Little by little, as the spider ran about trailing its drag line, additional threads were added around the egg sac so that a simple meshwork web was formed, much like that of long-legged *Pholeus*, and some theridiids of today. A continuation of this running about could give rise to the formation of a rough platform or sheet. In time, some of the spiders came to depend on the added protection of cracks in bark, fissures in rocks, and similar spots, so the eggs were deposited in these places where a retreat could be constructed.

Theridiid webs commonly appear to be just a tangled, three-dimensional mass of non-viscid threads, although viscid silk is used in the swathing film wrapped around the prey. But it is now known that some members of this family include a number of vertical threads that extend from the sheet to the substratum, and that these have at the lower end a number of viscid globules that can function in snaring ants and other ground insects. A reduction in the theridiid web is shown by *Episinus* of Europe. Here the sheet and irregular mesh are absent, and in effect the snare consists of only two of the "gum-footed" threads connected transversely above to present an H-shaped appearance. The spider hangs from the crosspiece by her hind legs and holds her front legs down along the vertical threads.

A still further reduction is that exhibited by a New Zealand theridiid, *Ulesanis pukeiwa.* Its web is made of a single thread, part of which is covered with viscid globules. The spider holds the thread taut, but jerks the line to entangle insects.

Finally, at what appears to be the end of the theridiid line of evolution, a few have divorced themselves completely from silk as a means of capturing prey. One example is *Euryopis,* which has no snare at all but lives under leaves and stones, having apparently reverted to the free-running hunting habits of its distant ancestors. Attention has already been called to the spitting spider, *Scytodes,* which has undoubtedly descended from ancestors similar to those of *Euryopis.*

We have seen how a sheet web may arise, and how in some cases the spider constructs a stopping maze above the sheet. If no tube or retreat is present, the spider lives on the sheet, hanging upside down from its lower surface. The typical web of this pattern is the product of various members of the family Linyphiidae. In *Pityohyphantes* the web is a flat sheet spun between the twigs of a shrub or tree. *Microlinyphia* makes a horizontal platform between stems of grass, and spins a stopping maze above it. *Prolinyphia* makes a filmy dome, and *Frontinella* spins a bowl-shaped web, with an additional flat sheet below and a stopping maze above. There is a tendency, especially with the smaller members of the family, for the web to show a reduction of the tangles above and below the sheet. This is especially the case in the majority of the Micryphantidae that have taken themselves and their webs down from the bushes to the ground, where they spin delicate, small sheets across depressions and irregular places.

The evolution of orb webs presents the greatest difficulty, and the question of how the elaborate method of constructing one was acquired by the ancestors of today's orb weavers is not easy to answer. The ecribellate orb weavers are very close morphologically to the Linyphiidae, and there is some similarity in the appearance of the webs too. Observations on the actual construction of orb webs indicate that the spiral is spun from the center out. This, then, is in effect the scaffold spiral, although the turns are much more closely set than is usual for such a structure. It has already been pointed out that in the webs of young *Uloborus,* as well as adult *Sybota,* the scaffold spiral is the only one in the snare. It has been suggested that dome webs of *Cytophora* and *Allepeira* might be regarded as a stage in the evolution of the orb. The snares of these spiders have a stopping maze, the radii are numerous, closely spaced and so branched that the interval between adjacent radii is only slightly nearer the edge than the center. These characteristics are considered primitive.

In the webs of the present-day silk spiders, those giants belonging to the genus *Nephila,* we still find branching radii; also, the scaffold spiral is left in the finished snare, the spider placing several turns of the viscid spiral between any two turns of the scaffold. A further characteristic that is considered primitive is the presence of the stopping maze in the *Nephila* snare.

When this scaffold is lost as a permanent part of the snare, but the stopping maze is retained, we advance to the type of web made by many of today's orb weavers, especially *Metepeira* and *Argiope. Metepeira* constructs its retreat within the maze, but *Argiope,* the common garden spider of America, is found in the center of her snare, which is generally furnished with a stabilimentum. Finally, the loss of the stopping maze, or at least its great reduction (the spider depending on the sheet alone) leads us to the typical snare made by the majority of higher orb weavers.

But even here we have modifications, leading to several lines of further evolution. One of the best known is that in which the spider almost always *(Zygiella),* or very commonly *(Neoscona),* spins an incomplete orb, omitting the viscid spiral threads from one sector. The spider makes loops and swings back and forth many times instead of going completely around the web when laying down the viscid line. Then the animal takes up its station in a retreat off the web itself, but remains connected to it by a signal thread extending from the hub and sometimes virtually bisecting the open sector.

A second modification occurs in the spider that bites out the ends of the radii where they meet at the center, so that an open hub results. Such snares are built by *Meta* and *Cercidia,* by the members of the Gasteracanthinae, and by the Tetragnathidae. The gasteracanthines make close-meshed webs with many radii and spirals, while the webs of the tetragnathids have few radii and spirals and are thus open-meshed. Moreover, the latter are horizontal, a condition considered more primitive than the vertical. A reduction has occurred with the genus *Pachygnatha,* where no web is spun, although the spiderlings build snares.

A third modification is seen in the Theridiosomatidae. The snare of the ray spider, *Theridiosoma,* was first described by H. C. McCook, one of the most outstanding American students of spiders during the nineteenth century. This spider spins a reasonably typical orb with a meshed hub and several turns of thread in the attachment zone. It then bites these out so that the finished web has no hub or attachment zone, and the radii are rearranged to radiate from, and converge upon, a point near the center. These rays, coming from four or five main divisions, join to form a trap line attached to some nearby twig. The spider stations itself on the rays at the center and pulls the orb into the shape of a cone or funnel with its greatly thickened legs. The web is held taut, but the spider releases the tension when an insect touches the snare; the resulting spring action ensnares the prey more firmly. Here, then, is still another example of the spring web, already discussed for *Hyptiotes, Miagrammopes,* and *Ulesanis.*

The fourth and final modification is the loss of the orb-making habit in certain genera of the subfamily Araneinae, and the substitution of other techniques for obtaining prey. Perhaps the most remarkable of these is the bolas-throwing behavior of *Mastophora* in America, *Dicrostichus* in Australia, and *Cladomelea* in South Africa. These spiders sit on, or suspend themselves from, a twig. They hold a thread, on the lower end of which is attached a sticky globule of silk, and fling it at passing insects. In *Cladomelea* the bolas is held by a hind leg and whirled rapidly in a horizontal plane. In the other two groups, the bolas is held by a front leg and is not whirled. Just how this behavior evolved from ancestors that undoubtedly were orb weavers, I cannot guess.

Another Australian spider, belonging to the genus *Celaenia,* is known to feed solely on night-flying moths, which it catches without the aid of either orb or bolas. From the published accounts it would appear that while the adult spiders wait in ambush and seize the moths by a lightning-quick movement of their legs, the spiderlings seem to construct snares. And so again we see, as with *Pachygnatha* and others, that the young revert to the habit abandoned by their adults. We find a parallelism for the structures built by the more primitive spiders (*Liphistius,* trapdoor spiders, diplurids, tarantulas, etc.) and the more advanced (so-called true spiders), and again between certain Cribellates and certain Ecribellates. We also find parallels among much more closely related members of a large family group, such as the members of the highly advanced orb weavers.

10

The Importance of Being Feverish

MATTHEW J. KLUGER

Most people associate a fever with the harmful effects of infection. In fact, pharmaceutical advertisements often give the impression that a fever is the cause of an illness, rather than a symptom, and that suppression of the fever is an effective treatment of the underlying infection. We are told to treat fevers with antipyretics, drugs designed to return our body temperature to normal, a treatment that has been an accepted part of medical practice at least since the ancient Romans began deriving aspirinlike salicylic acid from the bark of willow trees. That this body response, which has evolved over millions of years, might in fact be beneficial in killing infecting microorganisms is rarely implied in such advertisements.

The study of fever has always figured in medical history. Some 2400 years ago Hippocrates, who is considered one of the founders of Western medicine, attempted to explain the causes of the mysterious fever—malaria—raging through his country. Noting the correlation between local climatic conditions and initiation of the attacks of fever, he concluded that the weather was the cause of malaria (hence its name, which means "bad air"). His interpretation was erroneous, but based on the information available to him, this was a sound epidemiological approach.

Not only have physicians studied fever, they have also attempted to treat it with a wide assortment of remedies. Andromachos, the physician to Emperor Nero, proposed an instant fever cure-all made from more than sixty ingredients. His remedy was perhaps mild compared to others, which included fleas and the eyes of crabs, wolves, and snakes.

Not all physicians, however, advocated the abolition of fevers. Rufus of Ephesus, an anatomist-physiologist working in the first and second centuries A.D., believed that many non-febrile diseases, such as epilepsy, convulsions, and asthma, could be remedied by inducing a fever. This approach, subsequently called "fever therapy," is still a part of medical practice and has been used with varying degrees of success as a treatment for syphilis, gonorrhea, and some forms of cancer. Thus, medical practice, although most often attempting to suppress fever, does include two apparently opposed attitudes toward the phenomenon.

Our present understanding of the causes of fever was made possible only in relatively recent times. Not until the invention of the thermometer by Galileo Galilei in the late 1500s was it even possible to determine normal and febrile body temperatures. While the technology for the development of the thermometer had existed at least since the days of Hero of Alexandria in the first century B.C., it took the creative genius of Galileo to rediscover and appreciate this useful tool. Within a few years of Galileo's discovery, Sanctorius, his colleague at the University of Padua, used a crude thermometer, which was sensitive not only to changes in temperature but to barometric pressure as well (technically a "barothermograph"), to measure the "heat of persons in a fever." The use of temperature measurement as a diagnostic tool was thus initiated. But not until the development of the microscope by Galileo in the early seventeenth century, its subsequent refinement by Antony van Leeuwenhoek, and the later development of the germ theory of disease (largely the work of Louis Pasteur in the late nineteenth century) were scientists able to link the role of microorganisms with the onset of a fever. (Some fevers, of course, develop from noninfectious diseases, cancer for example, or from other causes such as severe allergy or injury, but our primary concern here is with fevers brought on by infection.)

While our knowledge concerning the course of fever has improved over the last 2400 years, we are still trying to determine its actual cause. By what mechanism do so many different pathogenic organisms all produce a similar febrile response? There are no definite answers, but our understanding is growing.

The primary area in our brain that receives information concerning temperature—both from the outside world and from deep body areas—is the hypothalamus. The hypothalamus serves as an area for the integration of all thermal information and also acts as a thermostat, regulating our body temperature at some prescribed level. When we are exposed to temperatures that are too high, the hypothalamus sends signals to our sweat glands to increase the output of sweat. (The heat required to evaporate sweat lowers the body temperature.) The hypothalamus also signals our metabolic machinery to lower our production of internal heat. Presumably through reflex pathways leading from the hypothalamus, we become conscious of the heat and we move to a cooler area. We also drink cold fluids, although this is a somewhat inefficient heat-loss mechanism.

Conversely, when we are exposed to the cold, information integrated in the hypothalamus leads to a cessation of sweating, an increase in metabolic heat production (shivering), and the conscious selection of a warmer area and warmer food and drink. Thus, the signals from the hypothalamus initiate both physiological and behavioral responses for body temperature regulation.

Fever begins with the presence of a foreign substance, say bacteria, in our tissues, which activates our leukocytes (white blood cells) to engulf or phagocytize the invaders. The bacteria needn't be alive since our leukocytes respond only to the cell walls of the bacteria, which contain the so-called endotoxin. In the process of ingesting the bacteria, the white blood cells produce a small protein called endogenous pyrogen. This pyrogenic, or fever-producing, material circulates throughout the body; some of it presumably enters the brain, where it causes an elevation of the hypothalamic thermostat.

Recently, scientists have speculated that endogenous pyrogen increases the production of special substances called prostaglandins; these in turn cause the hypothalamic set-point to rise. In any event, in response to the elevation in the hypothalamic thermostat, an animal—behaving as if it were exposed to the cold—elevates its body temperature and, as a result of subsequent thermoregulatory adjustment, develops a fever. Although antipyretics such as aspirin do not affect normal body temperature, they do lower the temperature during fever; the latest evidence indicates that antipyretics may reduce prostaglandin levels, which in turn return the hypothalamic thermostat to its normal setting.

Before exploring the phenomenon of fever, it is useful to distinguish between fever, a response to harmful bacteria, and hyperthermia, a response to exercise or heat exposure. Like a furnace whose thermostat is raised and which then works harder to produce more heat, during fever we act as if our hypothalamic thermostat is set at a higher level. Consequently, we actively drive our body temperature upward both by physiological means (such as shivering) and by behavioral mechanisms (perhaps wrapping ourselves with warm blankets and drinking hot tea).

In hyperthermia, however, the thermostat remains set at the same level, but the on/off switch fails. Consequently, the furnace overheats or, in this case, the body temperature elevates. Once we stop exercising or retreat from a hot environment, our body temperature returns to normal.

Although Hippocrates was speculating about the causes of fever more than two thousand years ago, even today we do not know whether it is beneficial or harmful to the organism suffering it. Some people have argued that the positive effect of fever therapy is evidence that fever itself is beneficial. In a patient undergoing fever therapy, however, the elevation in body temperature is both artificially induced and of greater height than that encountered during the normal course of an illness. Under normal conditions, the elevated body temperature does not directly destroy the infecting microorganisms. So the evidence from fever therapy clearly does not answer the question of fever's function during a normal infection, and we continue to be haunted by the following questions: Would a response such as fever, which is considered a universal response of warm-blooded animals, not serve some useful function? If fever were harmful to the host, would not selective pressures have led to its extinction?

To investigate the role of fever in disease, one could simply inject a population of animals with a suitable bacterium and allow half of the animals to develop the normal fever, while preventing the other half from developing the fever. The survival of the two populations could then be compared. If fever were beneficial, the population that developed it would have fewer deaths. Conversely, if it were harmful, the group that was prevented from developing a fever would have fewer deaths.

One difficulty presents itself, however. How could we prevent a fever from developing in a group of mammals exposed to a bacterial infection? The most obvious way is to simultaneously administer an antipyretic drug such as aspirin. Unfortunately, such an experiment would not give a definitive answer to the question. Since aspirin has numerous side effects, interpretation of any experiments using the drug would be difficult. Would any difference in mortality be due to the difference in body temperature, to the direct effects of the drug used to prevent the fever, or perhaps to a combination of both?

An alternate experimental design would entail manipulating the environmental temperature in such a way that one infected group was exposed to a comfortable environment and developed a normal fever, while the other was exposed to a cool environment that prevented the elevation of body temperature to the febrile level. Would this experiment be easier to interpret? I think not. In response to the cold, mammals initiate responses designed to prevent a fall in body temperature. Even if there were differences in the body temperatures of the two populations, there would undoubtedly also be differences in the amount of stress imposed on each group. The population exposed to the cold, for example, would have elevations in the levels of hormones partially responsible for the maintenance of normal body temperature.

What about cold-blooded organisms—the fishes, amphibians, and reptiles? To speak of fever in this group seems like a contradiction in terms. While birds and mammals (the so-called warm-blooded vertebrates) regulate their body temperature as a result of both physiological and behavioral adjustments, the reptiles, amphibians, and fishes regulate their body temperatures largely by behavioral adjustments. A turtle sitting on a log in the middle of a pond and a frog on a lily pad are familiar examples of cold-blooded animals raising their temperature by absorbing thermal radiation from the sun.

Over the past dozen or so years, laboratory investigations have shown that hypothalamic integration and control over the thermal responses of vertebrates, ranging from fishes to mammals, are similar. The primary difference between the cold-blooded (ectothermic) and warm-blooded (endothermic) vertebrates is the manner in which they regulate body temperature. Of the endothermic vertebrates, we also know that many, including birds, respond to infection with a fever.

What about the ectotherms? If they could develop a fever in response to a bacterial infection, we could design a definitive experiment that would answer the question of the function of fever. An ectotherm such as a lizard offers certain advantages for the experimental study of fever. In a natural setting, where there are large temperature differences from one microclimate to another (say from the shaded ground beneath a fern to a flat, exposed rock), a lizard can regulate its body temperature to within a narrow region. In a laboratory setting where the environmental temperature is relatively constant, the lizard's body temperature will remain at room temperature; thus, its body temperature can be easily maintained at any temperature by simply placing it in a chamber controlled at that temperature. This would enable us to inject populations of lizards with bacteria and study the role of temperature on their survival without the lizards attempting to alter their body temperatures physiologically.

We chose as our experimental animal the desert iguana *Dipsosaurus dorsalis,* a moderate-sized lizard about six inches long (excluding the tail) that adapts readily to laboratory conditions. The lizards were placed in a simulated desert environment, where the night temperature was a cold 55°F (12.8°C) and the daytime temperature ranged from 85° to more than 122° (29.4°–50°C), depending on the location (figure 10.1). Within this range the lizards were able to select their preferred temperature during the daytime. Using special thermometers placed in each lizard's rectum, we recorded their body temperature and found they selected a body temperature of about 100.4° to 102.2°F (38°–39°C).

Following infection with live *Aeromonas hydrophila,* a bacterium that causes red-leg infection in amphibians and reptiles, the lizards' body temperatures rose to between 104.0° and 107.6°F (40°–42°C). The lizards achieved this elevated body temperature only by selecting a site with a higher temperature a larger proportion of the time, not by increasing internal production of heat, as is largely the case with mammals.

We were now ready to test whether lizards infected with this bacterium and then placed in different constant environmental temperatures (that is, maintained at different body temperatures) would have different survival rates. Groups of infected lizards were placed in five different constant temperature chambers: at 93.2° and 96.8°F (34°–36°C), which correspond to low temperatures but are well within the lizards' normal range of exposure; at 100.4°, which is the normal body temperature of these animals; and finally at 104.0° and 107.6°, which correspond to low and moderate fevers respectively.

The results were striking. At the end of three days at these temperatures, 96 percent of the lizards maintained at a febrile temperature of 107.6° were alive, whereas at the a febrile temperature of 100.4° only 34 percent were alive. At 93.2° less than 10 percent survived. We later learned that the increased ability of lizards to survive at the elevated temperatures was due, not to differences in the growth patterns of the bacteria, but to some as yet unidentified increase in the lizards' defense mechanisms against the infecting bacteria.

Can these results be extrapolated to mammals? If specific aspects of the febrile response in birds and mammals could be shown to be similar, this would suggest a common origin of fever in these two groups. Mammals and birds evolved from primitive reptiles, so if mammalian fever originated in premammalian vertebrates (and did not evolve independently at a later time), then the function of fever might be similar in reptiles, birds, and mammals.

Hoping to increase our understanding of the evolution of fever, and perhaps its adaptive role, we decided to compare the febrile responses among the terrestrial vertebrates (reptiles, birds, and mammals).

In order to strengthen the case for a common origin of fever, we listed those characteristics of mammalian fever we felt should be found in the avian and reptilian classes. First, the reptile or bird should respond to an infection of live bacteria by developing a fever. Second, since the cell wall of the bacteria, not the live bacteria, contains the endotoxin that induces our own leukocytes to produce the fever-producing material (endogenous pyrogen), the reptile or bird should respond to an injection of dead bacteria by developing a fever. Third, antipyretic drugs should result in an attenuation of the fever. Lastly, in response to a bacterial infection, the reptiles or birds should produce endogenous pyrogen.

Our case was strengthened by our findings—all three classes of vertebrates developed a fever in response to live and dead bacteria, and the fever was attenuated by an antipyretic drug, thus satisfying the first three criteria. Still unresolved is the question of the development of endogenous pyrogen in reptiles and birds; investigations on this subject are under way.

On the basis of similarities of reptilian, avian, and mammalian fever, I believe that the mechanism responsible for the development of a fever in response to infection has existed for several hundred million years. I also believe that fever evolved as a mechanism to aid the host organism in surviving the attack of the infecting microorganisms. But how the elevation of body temperature leads to this enhanced body defense is completely unknown. Possibly, although there is no definitive laboratory evidence for this, several components of the defense mechanisms, including the phagocytic activity of the leukocytes or their ability to be rapidly mobilized are dependent on temperature. Perhaps, as has recently been suggested by Eugene Weinberg of Indiana University, a fever is beneficial because it leads to a reduction of trace metals, most notably iron, that are necessary for the growth of microorganisms, a phenomenon called "nutritional immunity." Future investigations in this area might show that our body has evolved a relatively simple, yet ingenious, system for fighting infection—the removal of some trace element that is critical for the growth of the pathogen. Ecologists have long been familiar with this phenomenon in terms of the requirements for the successful establishment of a species in an area and the role of "limiting factors" in the environment. In the case of long-term infection, however, or of poorly nourished individuals, this phenomenon can also lead to anemia and impaired functioning of the defense system.

In tracing the role of fever in disease, I believe that the comparative approach to fundamental biological questions, relying heavily on evolutionary biology, can lead us to answers we would not otherwise obtain.

Among the still unanswered questions are, What groups of mammals develop fever? How would a mammal that is capable of hibernating, such as a bat or a ground squirrel, respond to an infection? Can a fever be induced in an amphibian? a fish? If our speculation that fever is beneficial in mammals is correct, what causes the clearly dangerous high fevers that are occasionally encountered? Might fever be of positive value in some species and vestigial in others? By continuing to employ comparative techniques, we stand at the real beginning of a study first attempted 2400 years ago.

11

Goose Mates

FRED COOKE

The snow geese that sweep south out of the Canadian Arctic in large skeins each fall have long puzzled bird watchers, hunters, and scientists. For many years, the species was regarded as two: the snow goose, which, aside from black wing tips, has beautiful plumage as white as the snowy wastes where it breeds; and the blue goose, which has a white head but warm gray-brown feathers covering the rest of its body. Only in the last ten years have the two been recognized as representing different color phases of the same species. Interestingly, the plumage difference, or polymorphism, is primarily restricted to the Hudson Bay and Foxe Basin populations of the lesser snow goose (Anser caerulescens caerulescens).

In many species of birds, plumage differences are sexual. The plumage of the mallard drake, for example, is gaudy, while that of the female is cryptically colored. This is not the case with snow geese: both sexes occur in both color phases. This type of polymorphism is unusual in birds and is restricted to relatively few families, such as hawks, jaegers, and herons. One question for modern-day biologists is what significance such polymorphic differences may have. In a few cases, the differences have been shown to be under relatively simple genetic control, but in many others, the genetic basis has not been investigated.

One of the earliest insights into polymorphism in snow geese was made at Cornell more than twenty years ago by water-fowl biologist Graham Cooch, who now works with the Canadian Wildlife Service, and John Beardmore, a British population geneticist. Cooch had spent three lonely summers at Boas River, one of four snow goose colonies on Southhampton Island in Hudson Bay. Snow geese mate for life, and both members of a mated pair attend the nest, the female generally incubating and the gander standing guard. Cooch was thus able to record the plumage color of mated pairs and found to his surprise that he had evidence of color preference among his geese: birds usually paired with a bird of the same plumage color as themselves. Mixed pairs, while not unusual, were much less common than would have been expected if the birds had chosen mates at random with respect to the color of plumage. This phenomenon in known as positive assortative mating, that is, like choosing like.

I found Cooch's observation fascinating because it suggested that the birds were discriminating among potential mates on the basis of some readily identifiable visual cue. If this were so, the snow goose would provide a good subject for investigating the general question of how organisms go about choosing mates. Similar patterns of nonrandom mating have been reported for animals ranging from freshwater shrimps to Japanese quail, but many of these organisms are difficult to study under natural conditions. I decided to embark on a long-term study of snow geese to try to understand how mates are chosen among populations of animals in the wild.

In June 1968 I found myself camped on an isolated esker (a narrow ridge of materials deposited by a stream flowing in or under glacial ice) on the Manitoba shoreline of Hudson Bay, some 600 miles north of Winnipeg and 30 miles east of the little community of Churchill. I had come here because of the esker's proximity to the La Perouse Bay snow goose colony. Along the edge of the bay is an extensive arctic salt marsh dominated by goose grass and sedge. This marsh provides food for the goslings and their parents. Landward of the salt marsh is coastal tundra vegetation with small shrubby willows, lyme grass, and other assorted grasses and sedges. Most of the geese nest among this vegetation.

The La Perouse Bay colony is appealing in a number of ways, and I have returned every summer. Apart from being a relatively accessible and most exciting place to work (polar bears abound in late summer), the colony contains 4000 breeding pairs of snow geese, sufficient for a thorough analysis and yet not too many to handle. It also covers a small enough area so that most of the nests can be found every year and studied by a team of biology students. Our aim is to mark as many birds as possible and to observe these marked birds throughout their lifetimes. Only by knowing the individual birds, their genetic relationships, and their pair bonding can we ask detailed questions about mate choice.

Our marking methods are twofold. First, we place a small, numbered metal tag on each gosling the day it hatches. This tells us who the bird's siblings and parents are. The goslings soon leave the nest and move with their parents to the salt marshes, where they grow rapidly. At the same time, the adults molt their flight feathers and become flightless for three to four weeks. During this period, when the goslings are four to five weeks old, we carry out our second marking. With the aid of a helicopter and a crew of people on the ground, we round up large flocks of snow goose families. On realizing that they are surrounded, the geese become more docile and can be led into a net enclosure, which has been erected on a suitably dry part of the tundra. We can catch more than a thousand geese at a time this way, and each season we band up to eight thousand birds. The geese are given both a numbered U.S. Fish and Wildlife Service band and, for our own purposes, a plastic color-band with three letters on it. This second band can be seen at a distance and allows us to identify a bird without catching it. The band tells us the age and sex of the goose, when it was banded, and in some cases the identity of its parents. This last attribute is the most important in an investigation of mate choice.

When animals select a mate on the basis of physical and behavioral attributes, two types of choices may be important. First, certain attributes in some members of the opposite sex may be reproductively advantageous, contributing to the production of more offspring. Individuals with these qualities will often be chosen preferentially. Charles Darwin suggested that because birds that nest earlier in the breeding season can produce more offspring, females will compete for males that establish territories early in the season. On average, a female that obtains such a male will leave more offspring that one that does not.

A second possible consideration in choosing a mate, however, has to do not with this sort of general suitability (that is, recognized by all members of the opposite sex) but rather with suitability in relation to one's own attributes. When attributes are matched in this way, assortative mating results. I knew from the work of Cooch and Beardmore that there was evidence for assortative mating in snow geese, but observations of mating in wild populations can be misleading and I wanted to be certain that we interpreted correctly what we saw at La Perouse Bay. Chris Davies, a graduate student at Queen's University, and I have outlined five questions we feel must be addressed in studies of this kind. First, is there evidence of nonrandom mating in terms of some physical or behavioral character? If so, does this nonrandom mating necessarily imply mate choice? Again, if so, is the choice based on the character itself? Is there genetic variability in the population for the character? Finally, is there a selective advantage in making the correct choice? The final two questions are important in considerations of the evolutionary, as opposed to the immediate, consequences of mate choice.

At La Perouse Bay, we found a pattern of nonrandom mating very similar to that reported by Cooch and Beardmore. Of all the geese, approximately 28 percent were blue phase, and of all the pairs, 15 to 18 percent were mixed. According to statistical expectation, roughly 42 percent should have been mixed if pairing had been at random with respect to color. On the basis of data collected from the Boas River colony, Cooch and I had suggested in 1968 that goslings become imprinted on their parents and that later in life they choose to mate with a bird the same color as their parents. We reached this conclusion initially after looking at the genes responsible for plumage color. Homozygous blue-phase geese (those carrying two alleles for blue coloration) usually selected blue mates, whereas heterozygous blue geese (carrying a dominant allele for blue coloration and a recessive one for white coloration) often selected a white mate. We reasoned that homozygous geese must have had two blue-plumaged parents and that the heterozygous blue birds could have had one white-plumaged parent.

The genetic findings supported the hypothesis that parental color lay behind mate choice, but we had no direct evidence. Testing our hypothesis was the first objective of our La Perouse Bay study. After marking the goslings and recording the color of the parents, all we had to do was to wait and see which mates they brought back. A two- to three-year wait was necessary because geese do not nest in their first summer and only a fraction breed as two-year-olds. It took a number of years to build up an adequate sample size, but to date, more than 500 birds tagged as goslings have returned with mates. Of 353 birds with white parents, 90 percent returned with a white mate; of 103 birds with blue parents, 85 percent returned with a blue mate; and of 103 birds of mixed percentage—which we predicted would be willing to take either a white or a blue mate—60 percent came back with a white mate, 40 percent with a blue mate. Apart from a few exceptions, therefore, the findings were consistent with our hypothesis.

But are they really *choosing* mates? Without doubt there is an assortative mating pattern, but this does not necessarily mean that real choices are being made. For example, in *Homo sapiens,* married couples are usually of similar age. This could arise from one of two causes: either there is a preference for a mate of a similar age or there is no preference but potential mates of a similar age are more prevalent when the choice is made. In this latter case, a random selection is being made from a nonrandom sample of individuals. (The two causes need not be mutually exclusive; a preference may exist within a nonrandom sample.)

Determining whether preference or prevalence explained the nonrandom mating patterns in snow geese was difficult because pair formation takes place, not on the breeding grounds, but on the wintering grounds in the Gulf States or during migration. Birds from all the Hudson Bay and Foxe Basin colonies have a similar migration route and wintering distribution, so for these months of the year, the geese of La Perouse Bay are mixed with birds from most of the other arctic colonies. To complicate the situation even more, there is some geographical separation of the two color phases, perhaps reflecting their historical origins. Blue birds are more common in the eastern part of the winter range, predominating in Louisiana, while the white-phase birds are more common in Texas. Consequently, a blue (or white) bird is likely to be primarily in the company of other blue (or white) birds. This meant that if the birds do form pairs on the wintering grounds, we had to interpret our field data cautiously.

We decided to supplement our field data with experimental manipulation of the birds. We already had in captivity twelve mated pairs of snow geese of various color combinations. Each pair was placed in a separate pen and provided with six goslings hatched from eggs that we had collected in the Arctic and hatched in incubators. Some pure and some mixed foster families were created, but overall, there were equal numbers of white and blue birds. The goslings soon became imprinted on their foster parents, and after six weeks, all the families were released from the pens and allowed to mix freely in a large field. The results were unambiguous. All but two of the goslings from pure families chose a mate of the same color as their parents, while goslings from the mixed families mated at random with respect to color. These results, combined with our field observations, gave strong support for the view that geese do actively choose mates on the basis of plumage color and that the color they choose is influenced by the color of the birds in their own family. Both parental and sibling color were important, not just parental color as we had initially imagined. To further confirm our hypothesis, we raised more than a hundred goslings of both colors together. As we predicted, when these geese mated, they did so at random.

As a last check to be sure that the birds were really selecting mates by color and not by sound or smell or some other unknown, correlated character, we conducted a series of experiments that entertained and amused the local residents. We dyed some of the adult snow geese pink and kept others white. We then used them as foster parents and tried to find out whether goslings could discriminate between the two colors. As hoped, goslings with pink parents ran toward a strange pink bird in preference to a strange white bird, and the converse was true for goslings with white parents.

Another facet of this question that interests animal behaviorists is whether there are sexual differences in mate choice. Sociobiological theory suggests that since sperm are small and easier to produce than eggs, males are much less selective than females. While this is undoubtedly true in animals where polygyny is the rule, it may be less true in monogamous animals such as snow geese, which generally pair for life.

Unfortunately, a study of possible sexual differences in mate choice among snow geese is difficult to carry out on the breeding grounds because of the unequal numbers of banded males and females that return from the wintering grounds. Most birds have a strong urge, known as philopatry, to breed close to the area in which they were raised. Philopatry is thought to have evolved because a bird can survive and reproduce more effectively in a location with which it is familiar, and where it has learned the better feeding sites and areas where predators may be more prevalent. Among birds, males are usually more philopatric than females, but for snow geese—and probably all other waterfowl species—the reverse is true. At La Perouse Bay, for example, most of the banded birds that return to breed in their natal colony are females; very few males ever come back. Since the survival rates of the two sexes are approximately even, the males have presumably paired with females from other colonies and followed them back to their homes. While this behavior is interesting, it means that we seldom find out what color mates our male goslings wind up with. As the study has progressed, however, we have accumulated sufficient data on La Perouse Bay geese of both sexes to say that the males appear to be less discriminating than the females. Although both returning males and females are generally accompanied by a mate of the same color, males return with a mate of a different color more often than females do.

The differences in degree of philopatry result in an interesting pattern at La Perouse Bay: the number of mixed pairs where the female is white and the male blue considerably exceeds the number where the female is blue and the male white. The explanation for this is simply that La Perouse Bay (and Boas River, where Cooch and Beardmore observed the same pattern) is a predominantly white-phase colony. Since females are more likely than males to return to their natal colony to breed and since, in our colony, more females are white than blue, mixed pairs in the colony are more likely to consist of a white female and an immigrant blue male. In the predominantly blue-phase colonies, many of the immigrant males are white.

So far our studies have concentrated primarily on proximate questions. There is a long-term evolutionary question that we would ask as well—does choosing a mate of the same color confer any reproductive advantage? In 1935 Konrad Lorenz first suggested that a young bird learned what species it belonged to by becoming imprinted on its parents. He also proposed that if the bird subsequently chose a mate that looked like its parents, it would find a partner genetically similar to itself and with which it could breed successfully. Cross-fostering experiments, in which young birds of species A were raised with species B and vice versa, showed the general truth of this idea: the young birds chose mates of their foster parents' species. If imprinting, then, is a valuable mechanism for species recognition, what is the explanation for imprinting upon plumage color in a species that occurs in two distinct color phases? Is the recognition of polymorphism, as Pat Bateson, an ethologist from Cambridge, believes, a mechanism to allow an animal to choose a partner that is genetically even more similar to it and that thus might make a more suitable mate? Or does it serve some other function?

Organisms with color polymorphism have provided some of the best evidence of natural selection in action. A classic example is the peppered moth. Widely distributed throughout Britain, this moth comes in two forms, light and dark. In the latter half of the nineteenth century, the frequency of the dark, or melanistic, form increased dramatically in the industrialized parts of the country. The increase was shown to be a result of natural selection: in rural areas, the white moths are camouflaged on trees covered with light-colored lichens; in polluted areas, the lichens are killed, birds prey heavily on the now highly visible white moths, and the population shows an increase in the less conspicuous melanistic forms.

Early workers felt that similar changes were occurring among snow geese. Cooch believed that white birds were more vulnerable to human hunters and that blue-phase birds were on the increase. We have found no such increase within the La Perouse Bay colony. The overall survival and fecundity of the two plumage types are remarkably similar, suggesting that an evolutionary balance has been reached. Furthermore, the reproductive success of pure and mixed pairs is much the same. Is there then no evolutionary advantage to the color preference that we have documented?

Perhaps, we should consider the possibility that snow geese employ other attributes in addition to, not instead of, color to select a mate. Recently, we have found that large females tend to have large mates and small females, small mates. Although not proof that birds are selecting mates on the basis of size, this observation is suggestive. Also, in the case of size, unlike color, we can suggest a good evolutionary reason why they should do so. On average, pairs of large birds produce more eggs and more young than pairs of two small birds or of one small and one large bird. Considering size and color together, we can easily see potential conflicts for a bird that is looking for a mate. If mates are in short supply, the bird may have to choose between a mate of the right color but the wrong size and one of the right size but wrong color. How would the decision be made? This is just one of the questions we are now trying to answer. In the hope of unraveling this and other puzzles certain to arise, we expect to continue migrating north with the geese for many years to come.

12

Strategies of Reproduction

R. D. MARTIN

Some of the most basic structural and behavioral features of mammals concern reproduction, but reproductive patterns are seldom featured in discussion of mammalian evolution. This is curious because the mammals as a group earn their name from the universal possession of mammary glands and the associated occurrence of suckling behavior.

In fact, it has long been customary to divide the class Mammalia into three reproductive subclasses: the Prototheria (egg-laying monotremes), the Metatheria (marsupials), and the Eutheria (placental mammals). These three clear-cut divisions reflect the different degrees to which the fertilized egg develops within the mother before emergence. When all members of a particular mammalian group share a common pattern, this is a strong indication of an ancestral adaptation. All modern placental mammals, for example, exhibit a placental attachment of some kind during embryonic development. This permits the embryo to develop within the mother, so that live birth, or viviparity, occurs, rather than egg laying (oviparity). Therefore, a reasonable supposition would be that the ancestral placental mammals also exhibited this characteristic or something very close to it.

Reproductive characteristics are as fundamental to mammalian evolution as dental, cranial, or skeletal characters. Fossil jaws and teeth indicate that placental mammals had emerged approximately 100 million years ago. Because it also seems evident from reproductive patterns alone that all living placental mammals shared a common ancestor, we need to understand the evolution of reproduction in these mammals in order to interpret mammalian history.

One reason for the relative neglect of reproduction in recent work on mammalian evolution is that, at first sight, little can be learned from the fossil record. Paleontologists often imply that the only valid approach to the study of mammalian evolution lies in interpretation of hard parts preserved in the scanty fossil record. This approach, however, is also speculative, with the additional drawback that relatively few characteristics are preserved for consideration.

The exclusive advantage of a correctly interpreted fossil record is the establishment of an approximate time scale for evolution. A fossil fragment from a mammalian skeleton is nothing more than a geologic specimen until it is recognized as "mammalian" on the basis of comparison with some living mammalian species. Constructing hypotheses about the evolution of reproductive patterns in mammals through analysis of anatomical, ecological, and behavioral elements of present-day species is thus an equally legitimate undertaking.

The evolution of reproduction is of special interest in the case of the primates (prosimians, monkeys, apes, and man), since elaborate maternal care is central to many outstanding features of this particular mammalian group. To determine what factors might have influenced the evolution of primate reproduction, we must first identify characteristics that are typical of living primates.

A striking, universal attribute of nonhuman primates (excluding the tree shrews) is that their infants are physically well developed at birth. The eyes and the external ear channel are open at birth or soon afterward, the fur is well developed, and locomotion is sufficiently advanced for the young to clamber about.

This attribute of advanced physical development at birth is linked with several other characteristics to form a precocial complex. Following a long gestation period, primates typically have small litters (in most cases only one infant per birth), the subsequent lactation period is of long duration, sexual maturity is reached only after an extended period of development, and the maximum life span is considerable—at least a decade and increasing with body size.

The small litter size is matched by a small number of teats, with most primates having one pair. In addition, the brain size of both infants and adults is moderate to large, in comparison with other mammals.

Of course, most of these features—especially the gestation period and the brain size—vary systematically with body size, and such scaling effects must be taken into account in all comparisons. Nevertheless, it is obvious that all living primates exhibit the precocial complex, amounting in practice to a low reproductive turnover. Since virtually all primates have well-developed patterns of social organization in addition to, and dependent upon, their elaborate parental care, the overall trend in primate reproduction has clearly been toward quality rather than quantity. Fewer offspring are produced, but the few that are born are well cared for and are likely to live for a long time.

Adolf Portmann, who conducted a great deal of the systematic work on mammalian reproductive characteristics, found that mammals generally tend to fall into two distinct groups. Some, like the primates, are precocial; others exhibit an altricial complex. Altricial mammals typically build nests to shelter their off-spring and have relatively large litters of infants, which are poorly developed at birth (hairless, almost helpless young with their eyes and ears sealed by membranes). Gestation and lactation are of brief duration, sexual maturity occurs at an early age, and lifespans are typically short. Finally, altricial mammals have relatively small brains and lack elaborate social systems. Of course, there are intermediate groups and various exceptional cases that must be considered on their own merits, but it is notable that each mammalian order tends to exhibit predominantly one complex—either the precocial or the altricial. For example, insectivores (including tree shrews), rodents, and small-bodied carnivores are, in most instances, altricial; while primates, ungulates, cetaceans, and hyraxes are, almost without exception, precocial. Small-bodied mammals are generally altricial, whereas medium-sized to large mammals are generally precocial.

As observations of fact, all of these points are valuable, but is would be far more instructive to have a cohesive explanation for this divergence into two basic reproductive complexes in mammals. The common thread running through all of these characters is *reproductive turnover*. This is revealing, since natural selection amounts to differential survival of offspring. It is therefore highly likely that there are two essentially different reproductive strategies involved, in the sense that natural selection has generally favored two contrasting responses to the problems of population replacement and colonization of new habitats.

The divergence also applies to basic features of reproductive physiology; mammals with the precocial complex usually have long estrous cycles (periods of fertility) and exhibit spontaneous ovulation, whereas those with the altricial complex tend to have short estrous cycles with sharply defined times of maximum fertility and show some link between ovulation and the act of mating.

Interpreting the evolution of the estrous cycle is difficult, since under natural conditions most female mammals tend to exhibit a pregnancy cycle when they are in breeding condition. However, a short interval between successive ovulations in the absence of pregnancy really amounts to an adaptation for high reproductive turnover. If the interval between successive ovulations is brief, a female will soon become receptive again, should mating happen to be infertile on any one occasion.

The actual mechanism of ovulation is also somewhat difficult to interpret in terms of evolution. In any female mammal, one or more ripening follicles in each ovary will be mature when the female is receptive to mating. After the egg is released from the follicle (ovulation), the residual tissue of each ruptured follicle forms a *corpus luteum,* which produces progesterone to support the ensuing pregnancy.

In precocial mammals, the egg will typically erupt spontaneously at some fixed time, whether or not mating has taken place. The *corpus luteum* forms automatically and persists for some time before the next ovulation occurs. In altricial mammals, on the other hand, ovulation itself may depend on stimulation from the act of mating (with subsequent automatic formation of the *corpus luteum*) or ovulation may occur spontaneously but mating must occur before an active *corpus luteum* is formed. In either of these situations, the female will rapidly return to estrus if mating fails to take place, since no long-lasting *corpus luteum* forms and ovulation is not inhibited. Once again, the altricial mammals are clearly adapted for a high reproductive turnover.

The search for a general evolutionary explanation for the distinction between altricial and precocial mammals may perhaps be explained through the ecological concept of r-and K-selection proposed by Robert MacArthur and Edward Wilson. Briefly, their theory suggests that in an environment where there is little crowding and where food is relatively abundant, r-selection will favor those animals that harvest the most food and rear the largest families, that is, those with a higher population growth rate. This applies even if food is gathered wastefully, as natural selection will favor productivity in such a situation. By contrast, when there is intense competition for food and a given species population is consistently close to the carrying capacity of the environment, K-selection will favor those animals that can replace themselves with the lowest possible intake of food. Efficiency in the conversion of food into offspring is selected for, since in this case natural selection favors adaptive variations that serve to increase the carrying capacity of the environment.

According to this theory, if reproductive rates are the same, a large-bodied species consuming a given range of foods should reach the carrying capacity of those foods in a given environment more rapidly than a small-bodied species consuming the same foods. Therefore, efficient use of environmental resources is demanded of large-bodied species.

In principle, high rates of reproduction should predominate in situations where there are considerable fluctuations in the food supply, while more efficient, lower rates of reproduction should predominate where the food supply is relatively stable. Obviously, the two types of selection are opposite ends of a spectrum, and intermediate rates of reproduction must occur. However, there could well be a general two-way split that would match the distinction between altricial and precocial mammals.

Naturally, in discussing the course of mammalian evolution over the past 100 million years, one cannot begin to encompass long-term patterns or variation in ecological factors (for example, consistency of food supply). Nevertheless, some broad generalizations can be made.

First, it has been pointed out by MacArthur that tropical rain forest conditions would favor *K*-selection, while more seasonal and unpredictable environments would favor *r*-selection. In fact, the great majority of living primates live in tropical or subtropical forest conditions (this also applies to all well-established fossil primate species).

The nearly universal occurrence of the precocial complex among living primates indicates that the early primates must have undergone a considerable period of evolution under relatively stable conditions of food supply and competition. Adaptations during this period would have continued to influence subsequent evolutionary developments. One can suggest that the ancestral primates, which were most probably arboreal, were initially adapted for living in fairly stable tropical forest conditions with little climatic variation over the year. The precocial complex characteristic of living primates can therefore be regarded as an outcome of a reproductive strategy involving efficiency of exploitation of available food sources rather than high reproductive turnover. A relatively large brain size and a long period of mother-infant interaction, which permits social learning of such aspects of the environment as appropriate food sources, combined with the behavioral flexibility of a long-lived adult, would have contributed to the success of such a strategy.

The vast majority of living primates typically have only one infant at birth, which is usually carried on the fur of a parent (in most cases, the mother) for some time after birth, rather than being left in a nest. Here, the grasping big toes— adapted for arboreal locomotion on relatively fine supports—enable the infant to clasp the parent's fur. Riding along in this way, the infant is ideally situated to learn the techniques needed for future efficient exploitation of the habitat; and there is an additional result in that primate infants suckle "on demand" rather than on a schedule decided by the mother. Unlike many other mammals, primate mothers typically suckle their infants with great frequency, both during periods of activity and while sleeping. This intimate contact with the mother increases the infant's chances of survival and also provides the basis for developing social behaviors.

Most primates do not build nests. This is true of almost all species now inhabitating tropical rain forest areas. Although the great apes are known to construct sleeping platforms every night, the only primates to build nests at fixed sites are the bush babies and some of the smaller, nocturnal lemurs of Madagascar.

The majority of these small, nest-building prosimian species occur in dry forest areas. The dry forest bush babies and the dwarf and mouse lemurs are also the only prosimians that commonly have multiple litters (two to three infants) and produce more than one litter per year. Both features would be expected as adaptations to increased reproductive turnover, perhaps representing a secondary shift back to *r*-selection. This interpretation fits well with other evidence that the common ancestor of the bush babies and the mouse lemurs was adapted for relatively dry conditions. It might also explain the apparent paradox that bush babies and various noctural lemurs have precocial infants, just like other primates, yet construct nests.

It is apparent that the early primates became channeled toward low reproductive turnover. One question that remains is, were the earlier ancestral placental mammals altricial? as is generally assumed (mainly because modern altricial insectivores, such as the hedgehog, are widely taken as a model for ancestral mammals).

Portmann has pointed out that precocial mammalian species, such as the primates, seem to reflect an altricial ancestry in that their eyelids fuse together in development, as if in preparation for birth into a nest, and then reopen again at birth. Thus, although the primates can all be traced to an ancestor with precocial infants, the ancestral primate was probably derived from an earlier ancestral mammal with altricial infants.

It is also significant that the early mammals were probably small bodied and undoubtedly had relatively tiny brains, both of these features being typical of modern altricial mammals. Thus, it seems reasonable to conclude that the ancestral placental mammals were altricial, subjected primarily to selection for high productivity, and that one of the major shifts in the early stages of primate evolution was toward precocial patterns, under the influence of selection for efficiency in reproduction.

However, it should not be forgotten that all of these features are relative. A combination of viviparity and elaborate maternal care would itself be associated with reduction in reproductive turnover, when compared with a pattern of multiple egg laying. There was undoubtedly a gradual decrease in reproductive turnover as fishlike ancestors gave rise to the first terrestrial amphibians, as these gave rise to reptiles, and as the reptiles, in turn, gave rise to the mammals. Throughout the course of vertebrate evolution, there may have been a differential selection for more efficient and hence lower rates of reproduction in some species and higher reproductive turnover in others. Indeed, it is interesting to reflect that the concept of the phylogenetic scale, which has been so helpful when used carefully in discussing evolution, refers to a spectrum of increasing complexity in living organisms that is matched by a reduction in reproductive turnover. Man and the other primates lie at one extreme of this spectrum.

The establishment of the precocial complex among the primates provided much of the basis for human evolution, essentially because of original adaptations for low reproductive turnover. But this very feature of low fecundity has placed the primates, other than man, among the mammals most vulnerable to extinction. In the absence of natural checks, the human population is continuously increasing in numbers at the expense of the forest areas containing our primate relatives. On the one hand, removal of stable ecological conditions through technological advances has led to a population explosion in the most advanced primate species. On the other hand, disruption of the relatively stable forest conditions to which most other primate species are adapted must ultimately lead to the extinction of some of them. Primates as a group are not adapted for rapid colonization or recolonization, and the larger-bodied species are necessarily the most threatened because of their lower reproductive rates. Environmental stability was probably a key factor in primate evolution; the loss of such stability could be a key factor in primate extinction.

PART 3

SOCIAL ORGANIZATION

Introducing a section on social organization should be a piece of cake. After all, our own membership in a complex social species should enhance our appreciation of the benefits of social living. But like all other concepts in animal behavior, the term "social" is somewhat arbitrary and encompasses an enormous diversity of lifestyles. The most traditional view of social behavior generally includes only those animals living in relatively permanent groups. Regardless of whether the species are colonies of social insects, schools of fish, or troops of baboons, coordinated behavior clearly facilitates food acquisition, group defense, and mate selection. But along with the benefits come some disadvantages, such as increased competition for food and mates, which can easily increase the frequency of disruptive encounters.

If permanent groups represent one extreme of the range of animal interactions, passive aggregations constitute the other. For example, a group of fish thrown together by wave action into a common tide pool would not be considered a social unit. And even some active aggregations wouldn't fit the bill. For example, consider the many species of moths that are "lured" to a porch light on a summer night, or the variety of carrion-feeding animals that congregate at a carcass. In both cases, each individual is responding (at least initially) to an external stimulus. True social behavior, regardless of its level of complexity, is based upon communication among individuals of the species. The arbitrariness of the term "social" is also evidenced by the custom of restricting its use to behavioral interactions within a species. This eliminates interactions between predators and prey, parasites and hosts, and numerous other interspecific animal relationships. Finally, we also recognize that the very distinction between socially directed behavior and processes of individual maintenance and survival is frequently blurred. Thus an organism that is feeding, running from a predator, or even urinating may simultaneously be conveying information that can influence the behavior of other individuals.

The study of social behavior has recently shifted emphasis by focusing on the concepts and methodology that are subsumed under the heading "sociobiology." The principal goal of sociobiology is straightforward: to study the biological basis of social behavior by applying principles derived from population genetics and evolutionary theory. This approach is first illustrated in the article by Sherman and Morton on the breeding dynamics of Belding's ground squirrels. Next, in the paper on scrub jays by Fitzpatrick and Woolfenden, we are introduced to the surprising concept that reproductive potential can actually be enhanced by occasionally helping other birds to raise offspring. In Ewald's study of Anna's hummingbird in southern California we see how fluctuations of resources in an animal's external environment influence corresponding changes in its aggressive behavior.

The remaining five articles concentrate more on the diverse mechanisms that contribute to social organization. Topoff's paper on slavery in ants introduces what may seem like a paradoxical phenomenon: the occurrence of parasitic behavior in a group of social insects known primarily for their mutual cooperation. In her article on schooling behavior in fish, Shaw describes the visual communication that underlies schooling, and offers hypotheses about the adaptive value of swimming in such large groups. Next it's on to the lizards, where Simon explores the role of chemical communication in a group of reptiles whose social behavior was previously thought to be mediated almost exclusively by visual displays. Another dose of chemical communication is provided by Byers' study on peccaries, in which scent marking from the potent dorsal gland plays a primary role in integrating the movements and social interactions of the herd. This section concludes with a most unusual study by Curry-Lindahl, which documents the breakdown of social organization in lemmings, and attributes their behavioral pathology to endocrine malfunction during periods of stress.

13

Four Months of the Ground Squirrel

PAUL W. SHERMAN
and
MARTIN L. MORTON

Scene 1, spring

Snow still mantles the High Sierra of California in early May. At the summit of Tioga Pass, drifts are twelve feet deep. Suddenly, the pristine white expanse is disrupted by the appearance of a brown ground squirrel with short ears and tail. He sits quietly in the warm, dazzling sunlight, the first he has felt in eight months. In the next few days more males appear; they sit alertly near their burrows, paws folded on their chests, waiting and watching.

Scene 2, early summer

A ground squirrel creeps stealthily through meadow grasses and enters another squirrel's burrow. In seconds it emerges with a newborn pup, pink and hairless, hanging helplessly in its jaws. A quick bite to the head kills the youngster. Too late to save her offspring the mother arrives, attacks, and chases the killer away.

Scene 3, late summer

Staccato shrieks break the morning silence. Calling ground squirrels are standing on tiptoes, mouths gaping, chests heaving, and whiskers twitching. Not every individual calls, but every eye is riveted on a coyote that has just captured a mountain vole. As the predator departs, the calls die out, but the ground squirrels remain watchful, sitting stiffly at attention.

The performers in these intriguing dramas are Belding's ground squirrels (*Spermophilus beldingi*). For the ten summers from 1969 through 1978, we and our field assistants studied the ecology, physiology, and behavior of these animals. Now we are beginning to understand the biological significance of the scenes described above.

Belding's ground squirrels were named in 1888 by C. Hart Merriam, who received the first specimens from California ornithologist Lyman Belding. Today three subspecies are recognized. Collectively their range extends from central California north to Washington and east to central Nevada and Idaho. Historically they have been regarded mainly as agricultural pests, and until we began our studies, little was known about their natural history.

These ground squirrels are ideal study subjects for several reasons. They are large enough (8 to 10 inches long and weighing 7 to 10 ounces) to be seen easily, and they habituate readily to observers. Because they are diurnal, it is possible to observe all their aboveground activities. Finally, they are easily trapped and handled.

No physical features distinguish individuals, so we give ground squirrels permanent identities by attaching a tiny numbered tag to each ear. We have tagged 2681 animals, including all 747 young from 162 litters. We also mark ground squirrels for easy visual identification by applying hair dye to both sides of their bodies in unique combinations of letters, numbers, and symbols. These marks do not seem to affect the animals' behavior. From 1974 through 1978, we spent 4151 hours observing the behavior of our marked population.

Our research site was on top of Tioga Pass, which at 9941 feet is the highest highway pass in California. The study meadow, about a half mile long and a quarter mile wide, forms the bottom of a glacial valley; Pleistocene boulders and polished granitic outcrops give it relief. It is bordered on two sides by moraines and at the ends by Tioga Lake and Yosemite National Park. Between 250 and 300 Belding's ground squirrels live there each year. Individuals are active only during the summer, hibernating for the rest of the year (one of the longest hibernation periods of North American mammals).

Calendar dates of the ground squirrels' annual cycle vary with the duration of the winter (figure 13.1). From 1969 to 1978, dates of their emergence varied over six weeks, with animals coming out earliest in years of warm springs and light snowpacks. Each year adult males emerge first, tunneling through deep snow to appear one to two weeks before the females. As in scene 1, males remain near

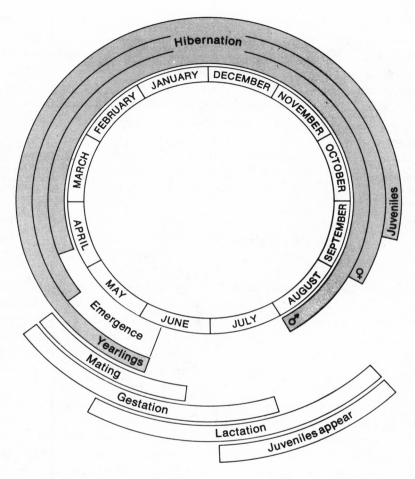

Figure 13.1. Annual cycle of Belding's Ground Squirrels at Tioga Pass, California.

their snow holes, waiting and watching for females. They cannot feed because no vegetation is available. Females do not emerge until snow has melted off the tops of the low rises beneath which they hibernate. In contrast to males, which usually hibernate alone, females frequently do so in groups; often such groups are composed of close relatives. In further contrast to males, which emerge physiologically ready to reproduce, females do not become sexually receptive until four to six days after emergence and they are receptive for only three to six hours on one day each year.

Belding's ground squirrels are unusual among terrestrial sciurids in that most precopulatory behavior and copulations occur above ground. When a female comes into estrus, males congregate near her. They do not defend territories or resources of value to females; instead they follow estrous females, attempting to mate and thwart mating attempts of rival males. Each year some males are highly polygynous, but many seldom or never mate. In 1978 for example, of twelve males under observation, three never mated and three more mated only once. The most polygynous three performed 46 of 65 copulations, and the most successful male mated 18 times with 11 different females (he alone accounted for 28 percent of all observed copulations). Heavy, old males (three or four years) win the most fights and so remain nearest to estrous females, enabling them to mate most often.

Completed copulations last ten to twelve minutes, but more than half the matings we observed were prematurely terminated by rival males. Perhaps partly as a result, most females mate more than once. Multiple mating may also enhance a female's reproductive success by assuring pregnancy and increasing the genetic diversity of her litter. Females mate an average of four times with three different males; one particularly promiscuous female copulated eleven times with eight different males. To determine which males sire litters we performed "paternity exclusion" analyses, using electrophoresis to identify polymorphic blood proteins of females, their mates, and offspring. We discovered that the majority of litters are multiply sired, that is, a mixture of full and half siblings, and that a female's first or first and second mates father most of the young.

After mating, females dig burrows and build grass-lined nests in them. Most nesting burrows are ten to fifteen feet long, one to two feet below ground, and have at least two surface openings. Females often change burrows, usually following flooding, infestations of ectoparasites, or predators' attempts to dig them out. One female changed burrows five times in a single summer. She dug three of the burrows herself and took over gopher burrows the other two times; each time she built a new nest, and in the process gathered 374 mouthfuls of dry grass.

Gestation lasts 23 to 25 days and the young are born in late June or early July. Each female has only one litter per season, and its size depends on maternal age; yearlings have three to four offspring; two- to five-year-olds have six to eight; and six- to eight-year-olds have three to four. Lactation lasts 25 to 28 days, so juveniles do not appear above ground until late July or early August. Weaning occurs coincident with, or soon after, emergence from the natal burrow. By the time the juveniles appear, some males have already gone into hibernation. They are followed by females in late September. Finally, when it begins to snow, young of the year begin their first eight-month period of dormancy.

Although primarily vegetarians, Belding's ground squirrels occasionally eat insects, birds' eggs, small mammals, and carrion. They are especially fond of flower heads and seeds; indeed, the generic name for ground, *Spermophilus,* means "seed loving." Sometimes their contortions to obtain seeds are comical. When harvesting grasses many times taller than themselves, the animals pull the stems down paw over paw, until the prize is reached. Then they feast, lying upside down, with all four paws clutching the stalk.

As fall approaches, the animals' consumption of vegetation and seeds increases dramatically. Unlike some hibernating sciurids, Belding's ground squirrels do not cache food for the winter. Instead they store energy as fat, and before hibernation their body weight nearly doubles. At the same time, the lipid content of their bodies increases approximately fifteenfold. During hibernation they use up about three-quarters of the stored fat, leaving a small reserve, which is used after their spring emergence. Thus, extensive overeating, leading to obesity, and the ability to metabolize fat during dormancy are adaptations for lengthy hibernation.

There is a sexual inequality in survivorship and longevity among Tioga Pass ground squirrels. Females usually live four to six years; at least four lived to be ten and two to be eleven. Males usually live only three to four years; the oldest male we saw was six. In addition to sexual difference in rates of senescence among polygamous creatures generally (members of the more polygamous sex die younger), injuries incurred during sexual combat contribute to differential male mortality. Fights over estrous females involve kicking, scratching, and biting and frequently result in lacerated throats, chests, and ears; broken toes, teeth, and tails; and dislocated shoulders. By the end of the mating period most males are hairless from mid-chest to mid-chin. Sometimes males are killed in fights; infected wounds hasten the demise of many more.

There are four additional mortality sources for Tioga Pass ground squirrels: weather, predators, infanticide, and automobiles (uncounted). Harsh winters are the major cause of death, with 54 to 93 percent of juveniles and 23 to 68 percent of adults perishing during hibernation. Adverse weather also causes mortality during the active season. For example, in April and May 1977 a 27-day-long storm occurred; it snowed every day and temperatures averaged well below freezing. About 60 percent of the animals active when the storm began died. They did not reenter hibernation and most presumably starved or froze after their fat reserves were exhausted.

The effects of the snowstorm illustrate dramatically the opposing selective forces that mold the ground squirrels' emergence and breeding schedule. Like other animals inhabiting regions with short and sharply defined growing seasons, they must emerge early and reproduce rapidly to ensure that their young reach appropriate body size before winter. But they must also remain in hibernation long enough to avoid spring storms and food shortages. In 1976, the snowpack was light and the spring weather temperate. Early emergers had plenty to eat, and their young were unusually large when they hibernated; many of them survived their first winter. Just a year later, due to the snowstorm, late emergers were favored.

The second mortality source is predation. Each year predators kill 4 to 11 percent of our study animals. Most deaths occur at night, when coyotes, badgers, or bears dig the ground squirrels out of their burrows. Weasels and coyotes are the major diurnal predators. Cooper's hawks, red-tailed hawks, and occasionally peregrine falcons, prairie falcons, and Clark's nutcrackers also attack ground squirrels, especially the young.

Infanticide, described in scene 2, is the major cause of mortality for nursing young and is perhaps the most intriguing. From 1974 to 1978, 8 percent of all unweaned juveniles were carried out of their burrows and killed by other ground squirrels. Adult females and yearling males are the most frequent killers. We did not see females kill their offspring or those of neighbors or close relatives, and females seldom ate their victims. When all their own young are lost to predators, females often migrate to safer sites and attempt to kill young there. If successful, they settle near their victims' burrows. Killing those juveniles likely to remain in preferred areas probably reduces future nest site competition for infancticidal females.

Like females, yearling males do not kill offspring of relatives. Unlike females, however, yearling males usually eat their victims and seldom settle near their victims' burrows. Yearling males also kill and eat mice, voles, and arthropods and feed more extensively on carrion than other age and sex classes, suggesting that meat is particularly important to them. Carnivory probably enhances the growth of yearling males, thereby increasing the likelihood of over-winter survival and copulatory success in their first mating attempts as two-year-olds. Over-winter survival for yearling males and copulatory success as two-year-olds are strongly dependent on body size. We suggest that the most likely factors underlying infanticide are competition for safe breeding places (females) and competition for mates (males).

Lactating females protect their young by defending territories surrounding their burrows against intrusions by unrelated conspecifics. They chase nonresident females and yearling males with particular vigor. The hypothesis that thwarting infanticide is the major function of territoriality is supported by three observations: (1) territorial defense ceases when young become capable of defending themselves against conspecifics (about the time they are weaned); (2) females that lose all their young to coyotes, badgers, or bears cease defense within a few days of the loss; and (3) females with the largest territories (those able to keep marauders farthest away from their burrows) suffer the fewest losses to intraspecific predators.

Perhaps the most interesting aspect of Belding's ground squirrel demography is their pattern of dispersal, which is also asymmetric by sex. Most females are sedentary from birth, spending their lives among near and distant female relatives. Granddaughters and great-granddaughters of females marked when our studies began will continue to occupy ancestral homesites. In contrast, juvenile males permanently disperse soon after weaning and establish burrows ten to twenty times farther from their natal burrows than their sisters (usually several hundred feet away). After mating, adult males also disperse, with the most polygynous moving farthest, sometimes as far as a quarter-mile. They remain in the area to which they immigrate and attempt to mate there the following spring. Thus males seldom interact with mates or mates' offspring and they do not behave parentally; as noted earlier, they often go into hibernation before the young appear above ground. Postweaning dispersal by juvenile males and postmating dispersal by adult males preclude incestuous matings. Advantages of avoiding consanguineous matings may have favored the sexual differences in dispersal.

The matrilocal population structure that results from a lack of female dispersal has led to nepotism. Because we have accurate genealogical records, we can compare competition and cooperation between mothers and daughters, littermate sisters, nonlittermate sisters (offspring of the same mother but different fathers, born in different years), aunts and nieces, first cousins, and grandmothers and granddaughters. We have discovered that the first three kin pairs are highly cooperative, and that cooperation among them decreases with decreasing genetic relationship.

There are four major manifestations of nepotism among females. First, they seldom chase or fight with their close kin when establishing nest burrows, so that among close relatives, females obtain residences with a minimum expenditure of energy and little danger of injury. Second, close relatives share portions of their territories and also join together in attacking potentially predatory conspecifics. They frequently chase trespassers away from temporarily unguarded burrows of close kin. Therefore, females with close relatives as neighbors lose fewer young to infanticide than females without kin. Third, females allow close relatives access to food and shelter within their territories; distant kin and nonrelatives are never permitted such trespassing liberties.

Finally, when predators approach, some ground squirrels give alarm calls. Calling is dangerous; we have seen more callers than noncallers attacked and killed by predators. Thus calling might be termed an "altruistic" behavior. From 1974 to 1978 we witnessed 119 natural interactions between ground squirrels and terrestrial predators, as described in scene 3. On these occasions individuals risking detection and capture by calling were usually old (4 to 8 + years), lactating, resident females with living offspring and sisters. Noncallers by contrast were mostly males, nonlactating, nonresident females, and females without living relatives. These age, sex, and kinship related differences in calling tendencies, coupled with the matri-local population structure, suggest that alarm calls function to alert relatives. Under this hypothesis, callers are trading the risk of exposure to predators for the safety and survival of dependent kin.

Belding's ground squirrels apparently recognize close relatives, and juveniles seem to learn who their kin are. Such learning takes place during social play near the natal burrow in the first few days the young are above ground (just before the litters begin to mingle). We know this because recognition "errors" occasionally occurred. At nightfall on their first day above ground, some juveniles entered a burrow other than their mother's (only one percent of youngsters make this mistake). The misplaced young remained in the nonfamilial burrows until their foster siblings came above ground; then they reemerged and played with the nonrelatives. The following year the misplaced youngsters cooperated with their foster mothers and sisters, with whom they had been socialized, but chased and fought with genetic kin as if they were unrelated.

Studies of the behavior of distant relatives suggest that there are limits to ground squirrel nepotism. Preliminary data indicate that aunts and nieces, first cousins, and grandmothers and granddaughters are uncooperative. They frequently chase and fight, eject each other from territories, fail to share territories or assist each other in territory defense, and refrain from giving alarm calls when they are solely in each other's presence. In short, distant relatives behave as if unrelated, and females with only distant kin as neighbors are no more successful in rearing young than females living among nonrelatives.

Why nepotism is limited to offspring, littermate sisters, and nonlittermate sisters is one of the enigmas that drew us back to Tioga Pass each spring. Perhaps distant kin cannot be recognized or else they are so rarely simultaneously alive that their cooperation has seldom been favored.

Our ten summers with Belding's ground squirrels taught us that to begin understanding the adaptive significance of behavior, long-term studies of individuals' ecological and social environments are necessary. We are struck by the relentlessness of reproductive competition among our animals, especially the males. At the same time we are fascinated by the degree of cooperation among close female relatives. Studies of individually marked animals throughout their lifetimes are being actively pursued around the world, and important new information about the natural history of insects, fishes, amphibians, birds, and mammals is emerging. We hope such work will proliferate and receive the level of acceptance and support we have experienced. By illuminating the significance of age and kinship in particular, studies such as ours contribute to an understanding not only of ground squirrels but also of the fundamental bases of social structure.

14

The Helpful Shall Inherit
the Scrub

JOHN W. FITZPATRICK
and
GLEN E. WOOLFENDEN

Deep in the interior of the beautiful Florida peninsula there exists a striking and distinctive native habitat that is rarely pictured on travel brochures. Visitors catch glimpses of it while cruising the central Florida highways, but most know it only as tiny patches of seemingly deserted scrubland scattered among lush orange groves, prairie pastures, golf courses, and housing developments. Perhaps the state's oldest habitat, and certainly one of the most endangered environments in North America, this the Florida oak scrub.

About fifteen bird species regularly breed in the Florida scrub, and one of these—the blue-and-gray Florida scrub jay—breeds nowhere else in the state. The species itself (*Aphelocoma coerulescens*) is not restricted to Florida; in fact, the Florida scrub jay represents a tiny, isolated population of a species that is widespread in western North America (where it is commonly encountered at lower elevations in shrubby habitats). But through careful observations during the past fifteen years, we have uncovered in the Florida population one of the most intriguing social systems of any bird; one very different from that of its western relatives. It involves cooperation within extended families, land acquisition and inheritance, faithful "marriages," and occasional "divorces." The origins of this unusual system are intimately tied to the past and present ecology of the Florida oak scrub.

The scrub grows only atop the rolling ridges of pure white sand that dot the peninsula, relict dunes that were deposited when much of Florida was inundated by shallow seas. Dominating this environment are several species of stunted oaks and a few other hardy shrubs. Complete with its own species of prickly pear cactus, this is a semiarid habitat, but not for lack of rainfall. The porous soil allows rainwater to run directly through to the limestone aquifer far below (supplying Miami with most of its drinking water). This leaves a parched and nutrient-poor soil, quite different from all others in Florida. Many species of plants and animals are completely restricted to these dunes. Unfortunately, the ridges also provide optimal conditions for citrus trees, one of Florida's largest industries. Thus, the scattered patches of oak scrub are disappearing, causing an entire community of distinctive species—including the Florida scrub jay—to become endangered.

We have been fortunate to be able to study a protected jay population at the Archbold Biological Station in south-central Florida. This ecological research facility is located near the southern tip of a sandy ridge that runs down the middle of the peninsula. At the Archbold Station as in some other places, the jays have become tame, affording us an unusually detailed glimpse into the private lives of a wild population of birds. By banding each jay with a unique combination of colored plastic leg bands, we have followed the births, movements, breeding histories, and deaths of hundreds of jays living in a beautiful, 1000-acre tract of prime oak scrub.

In 1969, the first year of study, Woolfenden discovered that, at four of the six scrub jay nests he found, more than two adult-plumaged jays were bringing food to the nestlings. This was odd because the young of most North American bird species leave their parents immediately upon reaching independence. We now know that Florida scrub jays have a social system, known as cooperative breeding, in which young birds remain at home for at least a year (and as long as seven years) and help their parents raise the offspring of subsequent years.

Although Florida scrub jays are in the minority, they are not the only birds to engage in cooperative breeding. Others include a variety of tropical jays, acorn and redcockaded woodpeckers, several species of South American wrens, some African bee-eaters, and many Asian babblers. The list is growing, and all such cases seem to present a nagging evolutionary paradox: given that animals evolved by competing for the maximum number of successful offspring left in future generations, why do some species regularly contain individuals that appear to forgo their own breeding efforts in order to assist the breeding of others? Many scientists believe that solving this paradox might shed light on the evolutionary origins of even more intricately cooperative societies, including that of our own species.

The evolution of cooperative breeding in the scrub jay becomes especially intriguing when one ponders the species as a whole. Throughout its vast population in the chaparral and scrublands of western North America, the scrub jay breeds like most other monogamous songbirds, nowhere showing helping behavior. As we shall see, this situation provides a clue to how cooperative breeding may evolve.

One of our first questions was whether these so-called helpers actually do help their parents. (Anyone who has been "helped" in the kitchen or workshop by a young, inexperienced child knows that not all helpers help.) Fortunately, our study population always contains pairs that breed without helpers, usually because they have no surviving offspring from recent years. This allows us to compare the performance of pairs that are assisted by helpers with those that are not.

Florida scrub jays normally raise a single brood of young annually, within a brief breeding season from March through May. Pairs with helpers fledge an average of 24 young jays per season, while those without helpers fledge only about 1.6 young. We have found that this difference does not simply reflect the natural variation in breeding success between experience birds (which are apt to have helpers) and inexperienced or traditionally unsuccessful birds (which are not). The difference persists even when the comparison includes only experienced birds. Indeed, even if we compare the performance of the same individual pairs, occupying the same piece of ground during years both with and without helpers, we find that they fledge more young when they have helpers. Without a doubt, Florida scrub jay helpers really do help their parents raise more young.

The scrub jays in Florida are nonmigratory, dwelling on permanent territories of about twenty acres, which the helpers aid in defending from neighboring families. Our study tract consists of about thirty such territories, within a much larger area of scrub containing hundreds more. In our thirty territories, where each of the 100 to 120 individuals is identifiable and relatively tame, we can investigate territorial skirmishes between neighbors and then plot the boundaries on an aerial photograph. By doing this every spring, we keep a continual record of the land occupied by each family from year to year.

Territorial boundaries are defended vigorously against intruders throughout the year, usually through aerial displays and raucous territorial calls, but occasionally by actual fighting. Jays will give up any other activity, including nest attendance, to defend their boundary against an intruding jay. Boundaries are narrow and clear-cut—often only as wide as a boot mark in the sand—although their exact locations may fluctuate between years. Every square inch of scrub in our study tract is occupied by a Florida scrub jay territory the year round.

Not only is the scrub constantly filled with territories, but the number of territories remains virtually constant year after year as well. In fact, the density of our population of breeding jays is more stable than that of any other population of land birds yet measured. This leads us to suspect that, unlike many other bird populations, this one is always near or at its maximum sustainable level.

These crowded conditions mean that compared with the number of jays capable of breeding at any one time, vacancies for these potential breeders are scarce and that, once obtained, land is jealously guarded. This results in a surplus population of nonbreeding jays that are, in a sense, waiting for their chance to breed. These nonbreeders are the jays that stay home for one or more years as helpers. Helping is not a permanent role, then, but rather a form of temporary employment before breeding.

Without exception, every territory contains a mated pair of jays. These, the owners of the territory, are faithful to one another as long as both are alive and well. Infrequently, a pair breaks up, generally when illness or injury causes one member to become less able to breed, defend the territory, or spot predators. The healthy member of such a pair may begin to court a new mate even while the old one lives. An impaired jay that recovers sometimes finds itself excluded from the territory it once owned. Some of these abandoned birds then pair in another territory; others roam singly until they disappear. Such "divorces" are relatively rare, however, occurring at a rate of about 5 percent. In the vast majority of cases, mated Florida scrub jays do not part until death.

Death comes swiftly, usually in the form of a hawk, snake, or bobcat. Resident Cooper's hawks, along with migrant sharp-shinned hawks and merlins, are the most dangerous of the raptors, while eastern coachwhip and indigo snakes are probably the two most common reptile predators. Here, another important role of helpers emerges. Among breeding jays with helpers, only about 15 percent fall to predators every year. Among those without helpers, the percentage rises to 23. Thus, by providing additional eyes to spot predators, helpers apparently enable their parents to live longer. Naturally, this benefit is reciprocal, so that the non-breeders also gain protection by living in a group.

Partly as a result of this cooperation, and partly through sheer luck, some breeders manage to escape death year after year. Our oldest jay of known age recently died, nearly fourteen years old. A few others are nearing that record now, and we have determined that in a healthy Florida scrub jay population, about one out of five breeders is at least ten years old. Such longevity is remarkable for a relatively small land bird, but it becomes a possibility only after a jay has established itself as a breeder. The story is more grim among the youngsters.

A young scrub jay must escape predation, starvation, and accident for eighteen days as an embryo inside an egg, another eighteen to twenty as a nestling, and then about two months as a dependent fledgling before reaching independence in late June or July. Mainly because of predators, only about one egg in four finishes up as a free-flying, brown-headed juvenile.

But the dangerous period is not over yet. The curious and clumsy juvenile jays take almost a year to attain the survival ability and food-finding efficiency of an adult. Only about one-third of all fledglings survive this first year out of the nest. We can only guess how much poorer their chances would be if they left home during this time. Here, then, is another reason for the extended family: it increases a young jay's chances of surviving during its long learning period in an environment that would otherwise be an even harsher one, filled with other jays zealously guarding their own ground. The final results are that only about one egg in thirteen produces a breeder, but nearly one-third of these live to ten years of age.

During their time at home, the young jays take their turns "on watch" as sentinels while other family members forage for food. They also participate in territorial defense, especially during the autumn and spring peaks in territorial activity when most boundary shifts take place. Most conspicuous of all, helpers bring food to the nestlings and fledglings being raised during the spring breeding seasons that follow their own birth.

Florida scrub jays build a bulky nest of twigs lined with fibers and tucked away only a few feet off the ground in a dense, often isolated shrub. Mated pairs neatly divide the labor of raising a brood, a common occurrence among permanently monogamous animals. The female performs all the incubation and brooding of young in the nest. (If a female breeder is killed during nesting, any eggs or young nestlings die of exposure the first night thereafter because males will not sit over them.) She, however, supplies little food, especially while the nestlings are small. As the three or four nestlings grow, the task of providing food can become formidable for her mate. Once again, enter the helpers.

A breeding male's workload can be decreased to less than half through the contributions made by even a single helper. Among helpers, yearling females do some feeding but males are more active, with older male helpers the most active providers of all. Some of our male breeders actually contributed only about 20 percent of the total food load brought to their own nests. This relief from the burden of brood rearing frees the breeding male to spend more time watching for nest predators, defending the territory, and keeping himself and his mate well fed.

So far, it may seem as if the breeders gain most of the advantages of the helper system. They produce more young, live longer, and work less hard, all simply by allowing offspring to help. Helpers gain a safe place to live while gaining experience, but do they benefit in any other ways? The answer appears to be a resounding yes. Unlike "helping" castes among social insects, scrub jay helpers are not destined to remain nonbreeders all their lives, and some of the most important benefits gained by helpers are related to the process by which they achieve breeding status.

Florida scrub jays become breeders through one of two routes. About half do so by replacing the roughly 20 percent of the breeders that die every year in the neighborhood. Male helpers rarely travel more than one or two territories to fill such breeding vacancies. Females, however, frequently disperse five or more territories away from home, and we have several records of females becoming breeders at least eighteen territories (about five miles) from their home ground. Far more mobile than males, females begin to search actively for breeding vacancies as early as their first spring, even while their brothers faithfully tend their parents' nest.

Thus, several important differences exist between the sexes: males tend to remain longer as helpers, are less active in searching the neighborhood for breeding vacancies, and are more likely to settle very near their natal territory. These differences are related to the second method of obtaining breeding space, a method open almost exclusively to males and perhaps a key to the whole system. We term the process territorial budding, but it could just as well be called by the familiar expression "inheriting the back forty."

Territorial budding appears to be an offshoot of one fundamental aspect of the Florida scrub jay's social system: bigger families occupy bigger territories. As a pair accumulates helpers over several years of successful breeding, the size of the family territory grows, usually from about 18 to about 26 acres. Older males cause the most pronounced increases in territorial size, probably because with age, they become more active and exploratory around certain boundaries. Since the oak scrub habitat has so little spare room, growth of one territory can occur only if others shrink. Less successful families lose ground to growing ones.

Following rapid territorial growth, one of the male helpers begins a more active defense of a segment of his family's territory. This male attracts a mate from outside his territory, usually one in her second or third year of life on a temporary dispersal foray. Pair formation can last many weeks, often interrupted when the female leaves to visit her home territory. The male often continues to help at his parents' nest during this courtship period. Slowly, however, the new pair's territory becomes better defined, and they begin to defend it together, even against the male's family. Typically, the immigrant female is the one to instigate the most intense skirmishes along the newly forming boundary because the male continues to treat those particular neighbors as family. His new mate becomes especially upset by the presence of her "mother-in-law," whom she knows only as a rival female. In this way, the two former helpers become breeders on a new territory carved out of land that the male helped acquire and defend while still a helper.

We begin to recognize the territorial budding system as an outright strategy when we uncovered one astonishing statistic: male helpers living in a territory with subordinate helpers are three times as likely to inherit land as those with no sibs at home. Their help at home thus pays off directly. In the safety of the home territory, and while acquiring skills necessary to become an effective and long-lived breeder, young scrub jays help to produce additional family members. As these younger siblings eventually contribute to the acquisition of additional land—a prerequisite for territorial budding—they substantially further the older male helpers' chances of becoming full-fledged breeders simply through inheritance. For these males, then, a fine way to become a breeder is never to leave home.

What's in it all for the females? With little chance of inheriting their breeding space (a few females actually have inherited some land), why should they bother to stay home as helpers? Their most important gain probably is during their first year, while they are still developing survival and breeding skills. Indeed, most females are less active helpers than males, and they usually leave home soon after their first birthday. This earlier departure does, however, result in a higher mortality among nonbreeding females than among their more sedentary brothers. The potential gains from long-term helping are apparently small enough to warrant their taking this extra risk in order to become breeders away from home.

In a nutshell, then, the rigors of a crowded environment outside the home territory appear to underlie the evolution of helpers in Florida scrub jays. This also explains why helpers have *not* evolved in western North America, where acceptable scrub jay habitat is widespread, from the semiarid deserts to the pine-oak woodlands in the mountains. Western scrub jays have not yet been studied in as great detail, but their social system is much simpler and more typical than that of the Florida population. Young birds disperse from their home territories during their first autumn, just a few months after fledging. During the winter, they join one another in wandering flocks of subadults of both sexes. At least a few succeed in settling into territories, obtaining a mate, and breeding the following spring. These sketchy facts support the notion that competition for space is less severe than in Florida, where usable habitat is restricted to a few relict patches of oak scrub. The quest for more solid evidence remains an exciting area for future work.

How might a complicated avian social system evolve from a more simple one? Fortunately for the purposes of answering this question, the Florida scrub jay system seems to represent a nearly perfect intermediate condition between the simple, noncooperative system typified by the western scrub jays and the advanced, highly cooperative system practiced by a closely related species, the Mexican, or gray-breasted, jay *(Aphelocoma ultramarina)*. This jay lives in mature oak and sycamore forests of northern Mexico and the extreme southwestern United States. Detailed information on the biology of this fascinating species is available thanks to long-term studies by Jerram L. Brown and his associates at the Southwestern Research Station in Arizona. According to Brown, competition for usable habitat also is severe in this species.

In the Mexican jay, offspring remain at home for several years before breeding, but group sizes are larger than in the Florida scrub jay. More important, two or three nests may be active simultaneously within the large, communally defended territory, a situation that is rare in our Florida population. Furthermore, virtually all jays in the group will feed nestlings at all nests, although each nest has a different principal pair, which still performs most of the activity there.

Brown has found that multiple nests in the Mexican jay occur when helpers form pair bonds with immigrant jays but stay within the natal unit. This is strikingly parallel to our finding that many Florida scrub jays pair with immigrants and remain on the natal ground. The difference, of course, is that in Florida a boundary develops between the new and the old pair, while no such boundary appears to form among Mexican jay units.

The Mexican jay system has made us especially interested in a few unusual cases among our Florida birds, where for several years no boundary developed between two breeding pairs. Some of these double families, as we call them, have become quite large. In one case, two brothers paired on a jointly held territory, much of which had belonged to their parents while they were helpers. Both bred successfully, and the offspring from each nest fed the nestlings and fledglings at both nests the following year. Unfortunately, one brother's mate died, so the situation changed. (He moved to an adjacent territory and paired with a female that his own son had been wooing. Just recently, the son won back this female from his aging father and paired with her after having helped her for several years.)

These exceptional cases of joint territories are functionally identical to Mexican jay social units, with two major exceptions. First, breeding Mexican jays often feed nestlings other than their own. This is a form of helping between breeders that we have never observed in Florida. Second, and more important, Florida scrub jays eventually do develop boundaries between breeding pairs, even when the breeders in one pair are related to the breeders in the other. This may take several years, allowing us a glimpse of a possible transitional stage toward a system like that of the Mexican jay, but single-pair territories remain the rule in Florida.

In the Mexican jay, then, some combination of environmental and population factors seems to have caused an evolutionary reduction in the aggression that typically develops between breeding pairs of jays and that results in the formation of boundaries. Perhaps the evolutionary pressures leading to reduced aggression, and ultimately to more complex extended families, are the same ones that led to the more elementary forms of cooperation practiced in Florida. The key limiting resource is usable land. If acceptable breeding space continues to become more limited over evolutionary time or if the jay population increases within the existing land, then competition for space can only increase. Holding on to land, and passing it between generations, is more difficult for small groups. The temporary double families of the Florida scrub jay might become increasingly successful and thus more common, until they are the rule instead of the exception in the population.

What better way to ensure that one's offspring, and *their* offspring, will successfully hold breeding space in a crowded environment than to help them acquire land, especially if in doing so one enjoys increased protection from predators and greater reproductive success? As an offspring, what better way to further one's own future success than to live in safety while gaining experience, knowledge of the community, and even access to land that one can inherit? If, by working as part of a larger unit, each member of the extended family becomes in its own way more successful, everybody gains. This, after all, is the very essence of cooperation.

15

The Hummingbird and the Calorie

PAUL W. EWALD

Summer brings dry heat to the hills of southern California, baking the moisture from the surface layers of soil. As the moisture vanishes in this chaparral community, so do most of the nectar-rich flowers so important to the survival and procreation of hummingbirds during winter and spring. During the summer months, only plants specially adapted for coping with the water shortage will bloom: scarlet monkey flowers cling to banks of receding rivulets and the drought-hardy tree tobacco blots up scanty pockets of underground moisture. But these flowers are insufficient to support large summer populations of hummingbirds—populations swollen with the year's crop of immatures, newly fledged from their nests.

Many hummingbirds abandon the chaparral as temperatures climb and supplies of nectar diminish. They migrate to higher elevations where wildflowers still bloom abundantly or to developed areas, where they can feed on cultivated flowers and at artificial feeders filled with sugar water.

Of the hummingbirds that remain in the chaparral many make local excursions to stands of California live oak and western sycamore that rise from canyon bottoms. Here hummingbirds can move from sunny to shady perches with little effort. Such easy access to a variety of microhabitats is especially important for the birds because their small size makes them particularly vulnerable to extreme temperatures.

Even in midsummer, early mornings are cool in the chaparral. Hummingbirds spend much of this time on sunny perches to reduce the high rate of heat loss caused by their small body size and high, constant body temperature. By reducing heat loss they can save valuable energy for other purposes. When isolation is strong and air temperature high, their body temperatures can rise quickly to lethal levels. So, as temperatures of sunny perches rise into the eighties, hummingbirds move to shady perches.

Chilling and overheating are not the only concerns of the hummingbirds. The limited supplies of nectar force them into keen competition for food. If flowers are sufficiently clustered and nectar rich, the rewards of evicting competing hummingbirds from these clusters may exceed the costs of defense, and certain hummingbirds will establish themselves as the "owners" of floral patches. Hummingbirds without such territories harvest nectar by intruding on owned floral patches or by feeding on flowers too sparsely distributed to be incorporated into a territory.

One denizen of the chaparral, the Anna's hummingbird *(Calypte anna)*, defends its territory by both ritualized communication and aerial attacks. The rituals are vocal or visual and fall into two major categories: announcement and threat. When food in a territory is not in immediate danger of being stolen, territorial males often become engrossed in "announcement song"—a recurring sequence of raspy triplets, a drawn-out creak, and two staccato chirps. The song broadcasts two messages: "I am present on this piece of property" and "I own it." While singing, the owner shifts its head from side to side, visually reinforcing its song with iridescent flashes of crimson.

The announcement song seems to function as a deterrent. After noticing the presence of an owner, intruders frequently avoid the territory without even attempting to steal food. Apparently, their foraging success is greater if they move on to undefended food sources or to territories whose owners are temporarily absent.

An intrusion on the territory, or the close approach of a would-be intruder, elicits a "chatter" from the owner. Like the buzz of a rattlesnake, the sound and meaning of the call threaten an attack.

A third line of defense—the gorget display—acts as a close-range threat. The resident bird fluffs up its feathers, creating an illusion of larger size, releases a barely audible, high-pitched peep, and vigorously throws its head from side to side. The neonlike flashes from the gorget are further accentuated by erection of the peripheral gorget feathers.

If these ritualized methods of defense fail to deter an invader, an aerial attack will usually ensue. If the intruder sees this approaching assault, it nearly always flees from the territory. However, if it does not see an attacking owner, actual contact may take place. When territorial ownership is at stake, such encounters will occasionally escalate into fierce combat: the contestants face off in midair, lunging with their bills and pummeling with their wings. Sometimes the pair may even fall to the ground during the tussle. I have never observed serious injury from such conflicts, but when I once mounted a deceased hummingbird in a perched position and placed it on a territory, the owner of the territory flew over for a closer look and then stabbed at the eyes of the mount with its bill.

When nectar is at stake, territorial conflicts are not restricted to competitors of the same species. Aerial attacks occur frequently between different species of hummingbirds and, occasionally, between entirely different phyla of animals. I once observed a territorial Anna's hummingbird unable to feed because it was repeatedly beaten back from its feeder by a paper wasp that was using it. However, hummingbirds are not always destined to defeat in such encounters; they occasionally attack and drive away bees that attempt to feed at their flowers.

Do hummingbirds use their repertoire of fighting skills selectively depending on the caloric gains and losses associated with each type of behavior? This question was investigated by reducing the amount of food in hummingbird territories on successive days. The technique was to provide a territory with a single food source—a specially designed feeder containing a specific concentration of sugar water. The amount of food available to the owner was reduced by adjusting a valve near the tip of the feeders. The results of this experiment showed that Anna's hummingbirds alter their defense tactics when the amount of food on the territory is reduced. Owners of rich territories streak from their perches like antimissile missiles to intercept invading hummingbirds. In such an encounter, the owner will frequently pursue a fleeing invader far beyond the territorial boundary.

In contrast, when territories are poor, owners shift from actual attacks to ritualized defense, using both chatter and gorget displays to a greater extent. Even the characteristics of actual attacks change; owners fly more slowly and attacks are of shorter duration, usually not continuing beyond the territorial boundary.

These alterations of defense tactics seem to be strategic modifications that increase an owner's prospects for survival. Actual attacks are more effective then rituals for expelling invaders, but they are more expensive energetically. Attacks are also probably associated with greater risks of injury. When territories are rich these costs are apparently offset by more exclusive control of food.

In poor territories, however, residents can keep the food sources well drained by frequent visits. If an intruder feeds, not only will it obtain little food but it should also be less inclined to return to that poor foraging spot. Under these conditions the best strategy appears to be ritualized defense against most intruders, and short chases when resources are in immediate peril—vigorous attacks being prohibitively costly for the meager benefits derived.

One purpose of this experiment was to discover the point at which a territory becomes so poor that an owner abandons defense completely. As the availability of food was reduced, owners left their territories more frequently and for greater lengths of time. Presumably, these departures were "grocery runs" through which owners fend off starvation by feeding on undefended food sources or by stealing drinks from other territories. Surprisingly, territories were defended even when the feeders provided no food.

This result suggests that territorial hummingbirds integrate both past and present information about territory quality when deciding whether to defend an area. This seems logical because energy obtained from a patch of flowers can fluctuate substantially even within a period of hours—some knowledge of how rich an area has been in the past could be helpful in deciding whether the area will be worth reserving for the future.

If this explanation is correct, areas that have always contained little or no food should not be defended. To test this hypothesis, feeders were placed on areas that had not previously contained any nectar-type food. As predicted, feeders of very low quality were not defended. Defense occurred only when the daily allowance of food from the feeders was more than about one-tenth of an owner's daily requirement.

In combination, these experiments show that hummingbirds are more likely to defend an extremely poor territory if it had previously been of high quality. But how much of the past influences the decision to defend a territory? A resident might need many days to determine how the overall productivity of a flower patch is changing. To gain insight into a territory owner's ability to incorporate information from the past, I conducted another experiment. Individuals were allowed to establish territories at feeders that provided unrestricted amounts of food for periods of one day to over a month. These feeders were then replaced with sham feeders that provided only water. When hummingbirds were allowed to use the rich feeders for one day, they abandoned defense of the sham feeders after approximately one hour. However, they defended shams for as long as two days when their ownership of rich feeders was increased to two weeks. Increasing their exposure to rich feeders beyond two weeks did not increase the subsequent defense of sham feeders. Thus, owners incorporate at least two weeks of past information in their decision to defend or abandon a territory. The biological mechanism could be one evaluation and memory or simply habit; determining which one remains to be solved.

These findings show how hummingbirds alter their defensive behavior as resources vary from day to day. But the dynamics of their territorial behavior are even more complex; minute-to-minute adjustments also occur. Owners are more likely to feed immediately before departing from the territory than at other times. These feedings reduce nectar loss to competitors that intrude on the territory during the owner's absence.

In addition, intruders encountered within the first few minutes after the owner has fed are less likely to be chased than those encountered later. By reducing the intensity of defense shortly after feeding, owners save on energetic expenditures and reduce their risk of sustaining injury from combat. Furthermore, little food is lost to intruders—even if an intruder feeds, it harvests little from the recently drained food sources.

These short-term modifications of territorial behavior are tinged with deception—they mimic the defense of poor-quality territories. An intruder encountering low food abundance on a territory and a passive (or absent) owner would not know whether this situation resulted from low productivity or a recent feeding by the owner. If the intruder knew that a recent foraging bout was the cause, its assessment of the territory's quality should be higher, and it should be more likely to return. The result for the owner would be increased costs of defense or losses of food.

These discoveries show that the repertoire of territorial behavior exhibited by the Anna's hummingbird is not merely a conglomerate of redundant messages. Rather, like a seasoned boxer, it uses its fighting skills selectively. "Knockout punches" are thrown when the expected gains outweigh the cost of escalation. But at other times, energy is conserved while the opponent wears itself down in its attempts to break through the owner's defenses. For hummingbirds, the reward is the elusive calorie, a key requisite for their survival.

Hummingbirds are brightly colored, and coloration is often used as a means of communication in animal societies. After a century of scientific debate, the exact information that is communicated by patterns of coloration remains enigmatic. Several social functions have been proposed: to attract sexual partners, to allow individuals to distinguish their species from other species, to aid in defense of resources, and to signal social status. The exaggerated gorget display used by Anna's hummingbirds is a clue that in this species bright color communicates information important for defense of resources.

Understanding the significance of such chromatic communication is especially challenging in hummingbird societies because many individuals do not possess a complete gorget. Anna's hummingbirds offer a typical example. When immature males fledge from the nest, they possess palled throats and foreheads, entirely devoid of red feathers. These iridescent feathers then appear sporadically until approximately seven months after fledging. At this time a dramatic alteration in the gorget region marks the entrance into adulthood: all feathers on the throat and forehead are replaced, transforming the patch iridescence of immatures into the immaculate gorget of the adult.

If bright coloration aids in territorial defense, bright individuals should require less effort than dull individuals to defend a territory. This idea was tested on

Anna's and black-chinned hummingbirds by simultaneously observing owners of adjacent territories. Each of these "next-door neighbors" was provided with identical feeders, but the owners differed in the amount of bright feathering in their gorgets.

Although artificial, such controlled techniques are important for field experiments because territorial behavior is strongly affected by the distribution and abundance of food, structure of vegetation, and time of day. Without this artificial control an experimenter would not know whether observed differences in territorial behavior were caused by differences between the owners or by differences in environmental variables.

In both species, bright owners spent less time and energy in aerial defense than dull owners, adults spent less than immatures, and bright immatures spent less than dull immatures. This result occurs because chases by bright owners are shorter and less frequent than those of dull owners. Since intruders frequently avoid a territory upon noticing the presence of an owner, bright coloration could decrease both the frequency and duration of chases simply by making owners more visible. Of course, the positive correlation between age and coloration yields an alternative explanation: the decreased defense expenditures could be caused entirely by increased experience.

If bright coloration is an asset for defending resources, another key question arises concerning patterns of coloration in hummingbird societies. Why do females and immature males usually possess little or no bright feathering on their throats and foreheads? There must be some compensating advantages to being dull.

One possibility is that dull individuals suffer less pressure from predators. Nesting females seem especially vulnerable; visual broadcasts from the mother during incubation and care of young could increase the risk of predation not only on the mother but also on the offspring. In most species of hummingbirds, these risks would influence only female coloration because males do not aid in incubation or raising of young.

Not surprisingly, females typically possess only a small patch of iridescent red feathers on their throats and none on their foreheads. In accordance with this less extensive coloration, females are territorial far less often than males. Female territoriality often occurs during the breeding season, but is usually limited to short chases at floral patches near the nest.

Anecdotal accounts yield a rather bizarre list of occasional predators on nonnesting hummingbirds: leopard frogs, bass, road runners, kestrels, and merlins. If brighter individuals suffered higher predation, bright coloration could be especially disadvantageous for immature birds. Their inexperience could make them not only more susceptible to predators but also less able to establish and maintain a territory. In this case, increased susceptibility to predators could offset the meager advantages of brightness gained through aggressive control of resources. The major weakness of this argument, however, is that predation on nonnesting humming-birds seems to be extremely rare.

Another possible advantage of dull coloration is greater success at stealing drinks from territory owners. Quantified observations support this idea. When owners are present, dull intruders gain entry at a greater rate than bright intruders. When owners are absent, there is no significant difference between the success of dull and bright intruders. Again, bright coloration may be causing these results by increasing the visibility of individuals. Bright intruders, being easier to detect, would not be able to feed as long before being expelled by an owner.

These results help explain why immatures are less brightly colored than adults. If the inexperience of immatures decreases their chances of owning a territory, they would benefit more than adults from the increased intrusion success that is associated with dull coloration. Not surprisingly, social structure during the nonbreeding season is related to age and sex; adult males tend to monopolize the best floral patches through territorial behavior. Females and immature males occasionally obtain territories, usually of low quality. More often, they survive on undefended resources or by stealing food from territory owners.

The explanations of dull immature plumage assume that the inexperience of immatures adversely affects their ability to defend territories and avoid predators. The gradual addition of bright feathers by immature males is consistent with this assumption; as immature males become more experienced, their chances of ob-taining a territory and their ability to avoid predators should increase. Hence the net benefit of bright coloration should increase with age.

Viewing animals under natural conditions can be crucial for understanding the importance of bright coloration. For example, the bright green plumage on the back of Anna's hummingbirds seems to function in a manner exactly opposite to the bright feathering in the gorget region. Rather than announcing the bird's presence, it tends to conceal a bird perching amidst green foliage. Such concealment could be important for avoiding predation or for allowing territorial intruders to avoid detection by owners.

The high energy existence of hummingbirds and the patchy distribution of their floral food sources lead to a social system centered around aggressive inter-actions. Our present knowledge provides a glimpse of how environmental condi-tions and antagonistic behavior shape the structure of this simple society. We do not yet fully understand these relationships for hummingbirds, yet the unknown answers and alternative explanations are far from discouraging. They generate the excitement that causes periodic migrations of curious scientists back to their field sites, while providing the blueprints for future research.

16

Invasion of the Booty Snatchers

HOWARD TOPOFF

Predators and parasites share a fundamental characteristic: they both survive at the expense of others. Nevertheless, most of us have very different perceptions of these two groups. Somehow we tend to accept predation as a natural part of the living world. Many predators, such as lions and other big cats, even command respect. We travel to remote regions of the world to see them, and we may even donate money to save them from extinction.

Parasites, on the other hand, evoke a different set of emotions, and disgust is probably near the top of most people's lists. We see parasitism as repulsive because it is often a long-drawn-out affair in which one animal latches on to another and slowly but systematically drains it of its life-supporting processes. Parasitism is an evolutionary ratchet, so to speak, because as parasites become increasingly specialized for one particular host, they often lose some of their own organic systems. When this occurs, there is no going back: the parasite is linked permanently to its host. And what is on the parasite membership list? Viruses, bacteria, liver flukes, tapeworms, and other associated organisms whose presence would never stop construction of a dam across the Chattahoochee River.

Given the bad press accorded parasitism, one might be surprised to learn that it has evolved in numerous species of ants—the animal group whose very name is synonymous with cooperation, sharing, and mutual dependence. Specifically, it occurs in eight ant genera from two distinct subfamilies. And just consider *Polyergus breviceps*, the western slavemaking ant, a species that I have been studying in the Chiricahua Mountains of southeastern Arizona. These ants are incapable of foraging for food, feeding their own broods or queen, or even maintaining their nests. Perhaps to compensate for these limitations, *Polyergus* ants have developed one extremely useful talent: the ability to get other species to perform these activities for them.

They accomplish this by conducting group raids on colonies of the related ant genus *Formica,* scattering the adult workers and queen, and snatching up the immature *Formica* pupae. Back at the *Polyergus* nest, some of the pupae are eaten; the remainder, however, are spared and reared. When these "adopted" *Formica* individuals emerge as adult workers, they promptly assume all responsibilities necessary to maintain the mixed-species nest. They forage all day for nectar and dead insects, regurgitate food to their *Polyergus* hosts, groom and feed the *Polyergus* larvae and queen, and even defend the nest against attack from other insects. Because they do not bring the *Formica* queen to their nest, the *Polyergus* must periodically replenish their work force by conducting slave raids throughout the summer. After several seasons of raiding, typical colonies of 3000 *Polyergus* ants may have more than 6000 *Formica* slaves working for them.

A *Polyergus* raid is a prime example of behavioral integration in a colony of social insects. During the morning and early afternoon on the day of a raid, the only visible activity at a *Polyergus* nest is the steady trickle of *Formica* foragers going to and from the colony. Starting about 3:00 P.M., however, several hundred to a few thousand *Polyergus* workers emerge from underground and begin to mill around the nest entrance. Within a short time, from one to six of the ants break away from the pack and wander off in search of target colonies of *Formica.* These are the *Polyergus* scouts, and our research shows that they are among the oldest and (presumably) the most experienced individuals in the colony. The scouts meander over the terrain, probing under rocks and leaf litter, until they encounter an appropriate nest. At our study site, "appropriate" means belonging to one species—*Formica gnava.* When a nest is located, the scout makes a beeline (or in this case an antline) back to her own colony.

Back at the mixed nest, the successful scout runs around excitedly in the milling ant swarm, contacting hundreds of *Polyergus* workers. This kind of tactile interaction, combined with a chemical trail, is a common way that ants of many species arouse large numbers of nest mates. In *Polyergus,* this recruitment behavior causes thousands of ants, both from the swarm milling around the entrance and from inside the nest, to assemble. The slave raid is on! The scout, with a troop of 2000 raiders, abruptly heads back toward the *Formica* nest. By marking scouts before the raid, we were able to observe them, and we noticed that they periodically darted in and out of the lead.

On a raid, the scout uses the sun to navigate back to the target colony. We found this out by performing a couple of experiments. First we suspended a wooden frame, covered with wax paper, 12 inches above an advancing swarm. The wax paper diffused the sunlight and interfered with the ants' ability to detect the sun's position. As soon as the raiding party passed under the wax paper canopy, the scout stopped, the workers scattered in all directions, and the raid was eventually aborted.

The results of the second test were even more dramatic. We suspended an opaque tarpaulin above another raiding swarm in late afternoon when the sun was low in the west. In addition to preventing the ants from seeing the sun or sky overhead, the tarp created a relatively dark area around the raiding party. As the slave raid moved from north to south, we waited until the scout entered this "twilight" zone. Standing on the eastern flank of the slave raid, we then used a large round mirror to relect the sun's image, thus creating the illusion that the sun was setting in the east. If the scout and raiders were indeed using the position of the sun as their compass, they would "conclude" that they were moving in the wrong direction. And sure enough, as soon as the ants encountered our artificial sun, they abruptly turned around and headed the other way.

The scout relies on optical orientation during the slave raid; the others use two cues for orientation. Raiders use the position of the sun to orient themselves, and they also follow the scout and rely on the chemical trail she deposits. Other ants that come upon the scout's trail also enter the advancing stream, the process snowballs, and the result is a very keyed-up ant army, a moving column about three feet wide and up to sixteen feet long. When we first discovered that *Polyergus* scouts navigate by the sun, we could not believe that this clue alone was behind their ability to find the exact location of the *Formica* nest. As it turns out, our skepticism was justified. When the scout arrives in the vicinity of the *Formica* colony, she stops advancing (and may also cease depositing the chemical). The other *Polyergus* raiders respond dramatically: they "screech" to a halt and start to move around the scout in ever widening circles. Indeed, their movements are reminiscent of their wanderings around their nest just before the raid. As they push outward, the raiders probe every leaf and rock crevice. If one of these ants encounters the *Formica* nest, she immediately recruits the entire swarm. So although the scout leads the raid swarm into the neighborhood of *Formica*, it is usually one of the 2000-plus raiders that finds the house.

When the *Polyergus* locate the target *Formica gnava* colony nest, they penetrate it immediately. Like other ants in the subfamily Formicinae, workers of *P. breviceps* have no stingers. Their principal weapons are their jaws, extremely sharp mandibles that can easily penetrate human flesh, not to mention the soft abdomen of a *Formica* worker. Formidable as the raiders are, however, they are not capricious killers. Instead of anatomical armaments, the *Polyergus* use chemical warfare. They spray the raided colony with a chemical (called a propaganda pheromone) that causes the *Formica* adult workers and queen to abandon their nest and scatter in all directions. The fleeing *Formica* attempt to escape with their brood, but most of the young get left behind in the melee. On a good day *Polyergus* ants are able to snatch almost 3000 *Formica* pupae, and when the raid is over, the *Polyergus* promptly take the booty back to their own nest. On the return trip there is no swarm, and no scout to act as leader. The raiders find their own way, using the sun and the residue of the chemical trail deposited on their way out. Upon their return, the *Polyergus* immediately turn over their booty to the adult *Formica* that are permanent residents of the mixed-species colony.

Once this transfer is completed it becomes obvious that the anthropomorphic designation "slave-making behavior" is woefully inadequate to describe what goes on between *Polyergus* and *Formica*. The functions that *Formica* perform for *Polyergus* probably result from a developmental process akin to imprinting, rather than from any kind of forced labor in the human sense. In many ant species, newly eclosed, or emerged, workers accept whatever odors they first encounter, even if these odors are different from those they would ordinarily experience in their nests. As a result of this developmental plasticity, researchers can create experimental colonies containing even unrelated genera of ants. In this case, *Formica* adults rear raided *Formica* pupae, together with the *Polyergus* brood, in the chemical and tactile environment of the mixed-species colony. As a result, when the *Formica* eclose into adult ants, they are familiar with, and accept, what would ordinarily be the foreign smell of *Polyergus* and go about their business in the same way as if they were "home" in their own colony. For example, adult *Formica* from the mixed colony forage on the same time schedule as their unenslaved *Formica* counterparts and even bring back the same kinds of foods. When we observed colonies in the laboratory we found that the *Formica* workers even feed the *Polyergus* adults, brood, and queen, in addition to feeding each other. Indeed, the *Polyergus* queen is so well attended by *Formica* workers that *Polyergus* workers are actually attacked when they attempt to approach her.

Our field observations of colony emigrations have convinced us that the *Formica* (and not *Polyergus*) workers also control the mixed-species nest. On the day following a slave raid, free-living colonies of *F. gnava* usually move to a new nesting site several feet away. During the past few years, we have also seen many emigrations of mixed-species colonies. These typically occur when one *Polyergus* colony stages a territorial raid on another, although emigrations may also be artificially set off by disturbing the nest. Regardless of the stimulus, all mixed-nest emigrations are organized and conducted exclusively by the *Formica*. Not a single

Polyergus ant has to move itself to the new nest. Instead, the queen, brood, and workers are all picked up and carried, one by one, to the new site.

By comparing the conditions of slavery in other ant species, we can hypothesize about how this form of social parasitism came to evolve in *Polyergus*.

Charles Darwin first suggested that slavery in ants might be derived from the more common behavior of predation. According to this hypothesis, the ancestral slave raiders were similar to present-day army ants, which conduct massive group raids on other ant species and carry away their immature forms. The principal difference, of course, is that army ants are strict carnivores and promptly consume all of their captured brood. The first serious step toward slavery occurs when (1) the predatory ants begin raiding closely related species and (2) some proportion of the raided pupae manage to avoid being eaten, surviving inside the predator's nest and eventually joining the work force. If these trends continue, the raiding species becomes a facultative slave maker, meaning it uses slaves but doesn't depend on them. This is exactly the stage of evolution represented by species such as *Formica sanguinea*. Here the worker caste is completely functional, so both master and slave participate in foraging, brood rearing, and all other tasks vital to colony life; the slaves are not yet essential.

Although Darwin's hypothesis is attractive, several scientists have recently suggested that territorial struggles, rather than food, may have been the prime evolutionary force leading to social parasitism. In several species of the ant genera *Leptothorax*, *Tetramorium*, and *Myrme cocystus*, fights often break out when ants from different colonies of the same species meet at a territorial boundary. After a prolonged battle, the stronger colony overruns the weaker one, captures its brood, and incorporates the emerging ants into its work force. We have seen neighboring colonies of *Polyergus breviceps* get into fights like this. Unlike the slave raids, these interactions result in many ants being killed. We observed that when both colonies were of approximately equal size, raiders were not able to penetrate the residents' nests. But in one very dramatic case, a large colony of *Polyergus* staged three territorial raids on a smaller colony over a period of two weeks and eventually wiped it out. Thus, according to this territorial hypothesis, the competition between colonies of Polyergus gave rise to raiding, and the behavior was somehow adapted for the related genus *Formica*.

Regardless of the origin of slave raiding, the fate of parasitic ants is similar to that of any other parasitic animal. As the parasite comes to rely more heavily on the host for nutrition and other basic needs, it eventually loses the ability to do things for itself. And when this happens, the species crosses the threshold and becomes an obligatory parasite. The four known species of *Polyergus* have all reached this stage of evolutionary specialization. Because the *Polyergus* workers do not forage or feed their young and queen, *Formica* slaves are truly essential for the colony's survival. So dependent are the *Polyergus*, it's almost as if the slave-maker has become the slave (and vice versa).

Like all successful parasites, however, *Polyergus* tenaciously retains one crucial function: the ability to produce new queens and males. For most ants, founding a colony is a straightforward activity. It begins with a courtship flight, during which males and females pair off in the air, come down to the ground, and mate. The newly mated queen then sheds her wings, excavates a small crevice, and lays the first small batch of eggs. We call this kind of colony founding "independent," because the queen tends this first tiny brood herself. The brood matures into adult daughters, which immediately take over the rearing of all future generations.

But *Polyergus breviceps* and other parasitic species are so dependent upon slaves that they can't form independent colonies. Instead of digging in on her own, the newly mated queen usually returns to her home colony and tags along on the next available slave raid, staying at the rear of the pack. When the raiders attack the target colony and drive out the *Formica* workers and queen, the *Polyergus* queen rushes in and takes up residence. Afterward, the *Formica* workers return to the nest and find this stranger in their midst, but instead of tearing her apart, they adopt her as their queen and reject their own. At present, we do not understand the communication process that allows this to happen. But a short time after the raid, colony life inside the *Formica* nest returns to "normal." The *Polyergus* queen lays her first batch of eggs, and the *Formica* rear the brood as if it were that of their own queen. A new generation of masters and slaves has been born, and the parasitic cycle is complete.

Polyergus is a good example of a social parasite, but it is by no means the ultimate example. This distinction belongs instead to the European ant *Teleutomyrmex schneideri*. In this species, the queens are fed by their hosts, *Tetramorium caespitum*. They contribute nothing to the host colony, and they produce only reproductive individuals. The worker caste is gone. In other words, *Teleutomyrmex* functions in some ways like a tapeworm. Somatic processes are reduced, but reproduction continues at full speed. Will this be the evolutionary fate of *Polyergus* too? Although it is impossible to predict how things will unfold, I am nevertheless tempted to speculate on the adaptations necessary for *Polyergus* to slide into the next phase of parasitism. Suppose, for example, that after a newly mated *Polyergus* queen took over an evacuated *Formica* nest, the resident *Formica* queen were allowed to return and resume egg laying. With such a readily available supply of foragers continually joining the work force, there would no longer be a need for slave raids. Without slave raids, who need workers? And parasites without workers are the ultimate parasites.

Most college-level surveys of the animal world start with protozoans and end with mammals. As a result, there is a tendency to equate increasing complexity with evolutionary success. The study of parasitism, whether physiological or behavioral, is a nice reminder that there is another side to the evolutionary coin: sometimes less is just as good.

17

Fish in Schools

EVELYN SHAW

Schooling behavior is almost as prevalent as feeding and reproducing among the approximately 20,000 species of fish. About 16,000 species school as juveniles and about 4000 of those continue to school throughout life. Schooling is found in a broad range of fish types, from the primitive, fragile anchovy to the advanced, powerful tuna. Schools of herring, cod, striped bass, and others make commercial fisheries feasible. Schooling is not only economically important, but also provides animal behaviorists with endless puzzles about behavioral mechanisms and adaptive advantages.

Many schools are large, containing thousands of fish, but any number can compose a school—from two to the millions of individuals found in a 17-mile-long spawning run of herring. There are no leaders, and fish at the forward edge may suddenly find themselves at the rear when the school reverses direction. School members are usually of the same species and the same approximate size. Smaller fish are not able to maintain the right speed: larger fish swim too far ahead.

The tendency to school is certainly genetically programmed, for it appears when fish have had little or no experience with each other. Fish can "recognize" their own kind from the very beginning of their lives. For example, *Menidia*, or silverside, fry, when they are about one-quarter inch long, drift aimlessly with other plankton and are exposed to a variety of vertebrate and invertebrate species, but when they reach one-half inch in length, they come together and form schools. The newly formed schools contain a single species only. This specificity is dramatically demonstrated in laboratory experiments when two related schooling species are mixed. After a brief period of trying one another out through approaches and body vibrations, the young fry separate in two species-specific groups. Clearly, at the very earliest stages, the fish are able to distinguish appropriate peers and are mutually attracted to one another. Indeed, mutual attraction is the primary criterion used for designating true schooling types.

But schools have other characteristics as well. When fish in a school move forward, they polarize, that is, they orient in parallel fashion—all individuals head in the same direction, swim at the same speed, and maintain fairly fixed distances from each other. When a school stops swimming, it depolarizes, but the fish remain together.

Parallel orientation and fish-to-fish spacing are distinctive qualities that give a school a characteristic three-dimensional structure—a geometry of its own. Fish in a school swim, not abreast, but in diagonal formation in a staggered pattern. When fish-to-fish spacing and diagonal positions change, the school looks quite different.

The most stunning feature of schooling is displayed when a school turns or changes direction. The fish then appear to be acting synchronously—turning together, increasing speed together, moving always in concert. This concurrent behavior entrances and baffles today's observers just as it mystified the scientists of the late nineteenth century, who postulated the existence of a "group mind" to explain the harmony of thousands of fish acting as one. The group mind concept had a short life, however, and twentieth-century scientists are discovering the biological mechanisms responsible for the synchrony and characteristic organization of schools.

In the 1920s experimenters identified vision as the sensory modality critical to schooling. Sightless fish do not approach other fish and consequently cannot school. Even though schooling fish must be able to see, they do not need to distinguish much detail for the school to maintain its coherence. A crescent moon, the stars, or phosphorescent organisms that cling to the skin of some species, all give sufficient light to enable schooling to continue. In one laboratory experiment, a school of jack mackerels maintained cohesion even when the light was so dim that it required ten minutes of adjustment before the observer could see the school. In the total absence of light, schools will disperse. Fish observed in the dark through a "snooperscope" were found to be swimming randomly inside a tank, seemingly unaware of each other's presence.

In addition to the visual sense, which is not only primary but which probably also serves as the major pathway of communication, other senses such as the olfactory and auditory may function in schooling. It is well known that fish have a fine ability to discriminate between even the faintest odors. It seems unlikely, however, that changes in the speed and direction of a school are communicated through a schooling pheromone, a chemical substance that conveys intraspecies information, but olfactory cues may operate in species discrimination, that is, in the recognition of similar types. Communication of changes in speed and direction through the production of sound is also unlikely. Schooling fish tend to be almost noiseless. Several experimenters have found that at night, when visual references are unavailable, some species produce sounds that may serve to keep the group together. These sounds are not detected, however, in the daytime in an actively moving school.

All fishes possess a special sensory system, called the lateral line, that detects changes in the movement and pressure of surrounding water. In many species the lateral line consists of a series of canals crisscrossing the head in a characteristic species-typical pattern. In addition, most fishes have a single canal that courses down the lateral body wall on each side. The basic sensor of the lateral line is the neuromast, a cluster of innervated hair cells capped by a gelatinous sheath known as a cupula. Neuromasts are spaced regularly along the canal. When water is displaced around a fish, the movement is detected by a neuromast through the bending of the cupula. Fish can see changes in the spacing, speed, and direction of schools, but because of the sensitivity of the lateral line system to movements and pressure changes in the surrounding water, it may serve as a prime source of information.

Exactly how the lateral line functions in schooling remains a mystery, but that it does function is suggested by the following experiments on tuna conducted by Phyllis Cahn, an acoustic physiologist at Long Island University, New York, and on jacks by the author. When jacks were artificially separated from each other by a transparent partition that blocked reception in the lateral line, the fish moved closer together than in a normal school and fish-to-fish spacing was markedly reduced. Tuna, on the other hand, will spread farther apart—from two to four times their normal distance—under similar experimental conditions. The loss of information from water displacements appears to upset the characteristic spacing maintained by schooling fish, and sight alone evidently cannot compensate for the lack of input derived from water movements.

In another experiment, John R. Hunter and Jon Van Olst, of the National Oceanic and Atmospheric Administration (NOAA) at La Jolla, California, showed that spacing and fish length are related. When they are young, jacks, mackerels, silversides, and anchovies swim in loosely structured schools, and the spaces between the fish measure about three to four body lengths. When the fish grow to three or four inches, the spacing shrinks to half a body length for all four species. Spacing of half a body length is also found in other species, which suggests that this may be the optimum distance for obtaining the most precise water displacement information through the lateral line sensors.

A last point about schooling and the senses concerns the relationship between speed of response and the visual system. Hunter found that a jack mackerel reacts more quickly to changes in a neighbor's activities if it sees the neighbor in certain areas of its visual field. By photographing one tethered fish and five freely swimming fish, he determined that the responding fish reacts more quickly to a neighbor's behavioral change if the neighbor is either directly ahead and can be seen with both eyes simultaneously or if the neighbor is at such an angle alongside that its image fills the entire visual field of one of the responder's eyes. The reaction time of the responder fish to changes in direction or speed is a mere 0.15 to 0.20 seconds, hardly more than a blink of an observer's eye. Indeed, the reaction time is so short that, to the unaided eye, an entire school of fish may seem to turn simultaneously in response to a change initiated by only a few individuals. This ability react quickly is highly useful for survival.

All evidence indicates that it is beneficial for fish to be members of a school. Fish in schools swim for longer periods, cover greater distances, and tolerate colder temperatures than comparable fish swimming alone. Diverse experiments show that groups of schooling fish consume less oxygen per fish than an individual school member swimming alone. Goldfish in a group tolerate higher doses of toxins per fish than when alone, sunfish learn the pathways of a maze more quickly in groups than as singles, and groups of carp avoid a moving net more successfully than loners. An individual fish trained with a group and then removed regresses to poorer scores. The advantages of group life seemingly cannot be maintained by single fish.

From one point of view, schooling might be considered a form of social behavior in which the actions of a few benefit the many. For example, when a predator swoops out of the sky and plunges into a school of prey, the fish in the predator's immediate vicinity, reacting swiftly, alter their swimming course and speed in order to escape. Through visual signals and shifts in the water pattern, these changes are rapidly communicated to other fish in the school. A wave of disturbance will sweep through the school until the whole group diverges from its original path. The total elapsed time is perhaps half a second. In this case, the response of a few has benefited the entire group, and the school has escaped from the predator. A few individuals may have been consumed, but their loss is insignificant to the group.

Each member of a school can be thought of as a scanner of its own environment. In a randomly chosen school there may be hundreds, thousands, or even millions of scanners. Fish at the edges of the school become the "eyes" of the fish in the center, thereby enabling any fish to "see" beyond the limits of its actual field of vision. It is not necessary for fish in the center of the school to be constantly on the alert to potential danger; all they need do is respond rapidly to changes in their neighbors' behavior. The same holds true when a food supply is sighted; then too, the fish in the center are informed. When tuna sight prey, for example, dark bands form on their flanks, signaling other fish that food is nearby. In this manner, many fish can feast even though only a few have discovered the food.

Schooling has other attributes, depending on whether the school consists of prey fish or predators. A theoretical model, based on chance encounters during a given period of time within a large area, indicates that school prey should maintain a highly compact pattern and move very slowly in order to reduce the probability of an encounter with a group of predators. School predators, on the other hand, require encounters. The likelihood of detecting prey is theoretically enhanced if the predators move rapidly in looser schools so as to cover a large area. In fact, both types of schooling have been observed in nature.

When an encounter occurs, it is, of course, disadvantageous to be the prey, but single fish prey are at an even greater disadvantage than school prey. This was demonstrated by a Russian investigator who introduced young coalfish, alone and in schools of 25 to 35, to predatory cod. Two and a half minutes elapsed before a cod consumed a single coalfish school member, but only a half minute elapsed before the cod consumed a coalfish introduced singly. The cod was unable to concentrate on any one fish in the school and was distracted again and again from its target fish, switching its pursuit from one individual to another until it finally focused on its victim. The school created a "confusion" effect, which provided time for all the fish to escape from pursuit except, ultimately, the victim.

As mentioned previously, schools generally contain only one species. Experiments confirm that there are advantages to acting and looking like your neighbor. Predators tend to select individuals that differ from their companions; they pursue the weak, the slow, the conspicuous. One experimenter marked prey fish with fluorescent dyes and found that the glowing fish provided brilliant targets for predators. Their shining hour came to a quick end. And, in nature, the flash of silver from the gill cover of an anchovy as it feeds pinpoints the fish as prey despite its membership in a school. Because a feeding anchovy no longer resembles its neighbors, it can be picked off.

Even regular spacing of fish in schools works to the advantage of the prey. Edmund Hobson of NOAA watched a grouper intently watching a school of herring. The grouper attacked only three times in two hours—when the school dispersed in response to a diving pelican. Herring that strayed too far from the aggregation were quickly gulped into the grouper's wide mouth. Dispersion also occurs when two prey schools encounter each other and their respective geometric patterns are temporarily disrupted. Before the schools can reassemble, individual fish are particularly vulnerable to predators.

Another survival feature of school geometry applies to prey and predators alike. Studies show that in fish-to-fish spacing, both the distances between individuals and their diagonal positions are important factors in the conservation of energy by a school, whether it be prey or predator. A diamond school pattern maximizes the energy available to the member fishes. Fish propel themselves forward by tail thrusts. As the tail sweeps from side to side, it creates vortices, or small whirls of water, in the fish's wake. Depending on its position in the school, one fish can coast on the vortices produced by another in front of it, thus spending less energy on swimming. Fish change position within the school with sufficient frequency to guarantee that no one fish must continuously work harder than any other. In this manner, one fish can utilize the energy expended by another—energy that would otherwise be dissipated and lost—and the overall effect benefits the entire school.

It is apparent that many good things happen to fish when they take up a schooling life. They have constant companionship and readily available partners during the reproductive season, they make maximum use of available food, conserve their energy, and derive such social benefits as protection from predators and enhanced learning ability.

18

Masters of the Tongue Flick

CAROL A. SIMON

Many lizards, like many snakes, often stick out their tongues. Some simply flick them into the air, while others touch the ground with their dexterous appendages. This behavior fits in well with some popular stereotypes of things reptilian: the flicking tongue connotes mystery, danger, and sinister forces. The truth about tongue flicking is much more mundane. Several years ago my colleagues and I began a series of field and laboratory studies aimed at answering two fundamental questions: What do lizard tongue extrusions accomplish, and why do some species stick out their tongues continually while others do so only once or twice an hour, if that frequently?

Several obvious answers come to mind. Lizards certainly use their tongues to manipulate food and water, but they also flick their tongues into the air, and those that touch their tongues to the ground do so at the same rate whether or not food and water are present. Tasting is a possible reason, but most lizards have few taste buds and these are usually located at the back of the throat. Taste is believed to be a relatively unimportant sense in lizards.

Lizards and snakes are closely related, and since many snakes do a great deal of tongue flicking, some knowledge of how they use their tongues is helpful in determining why lizards stick out theirs. When a snake or lizard flicks out its tongue, airborne chemicals adhere to its moist surface, and as the tongue is retracted into the mouth, it slides across the palate, where small slits open to the vomeronasal, or Jacobson's, organs. These paired chemosensory organs are completely hidden by the palate, and the molecules brought in by the tongue are swept by cilia through the slits and into the organs (figures 18.1, 18.2). A nerve, the vomeronasal, leads directly from the Jacobson's organs to the brain, so that information concerning the nature of the chemicals can then be processed by the central nervous system. Lizards also have paired Jacobson's organs, which

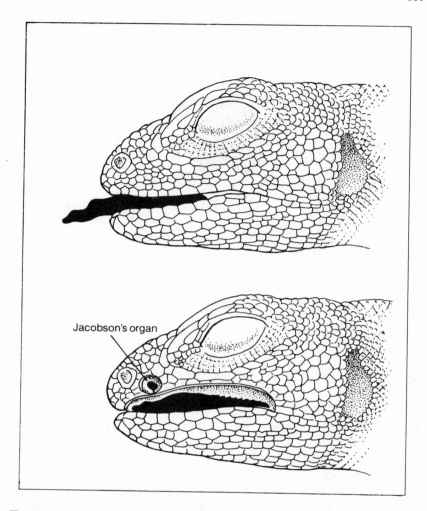

Figure 18.1. Desert Iguana *(Dipsoaurus dorsalis)*, vomeronasal system. *(drawing by Frances Zweifel).*

are well developed in some species but poorly developed in others. Although few physiological studies have been done on these organs in lizards, they seem to function similarly to those found in snakes. (In both snakes and lizards smell is a completely separate process.)

Studies suggest many possible functions for the lizard vomeronasal system. Consider the Gila monster, the only poisonous lizard in the United States. Gila

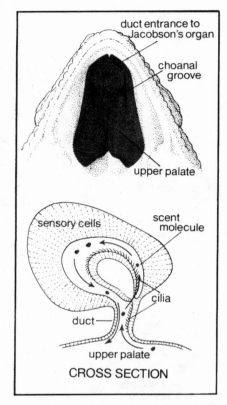

Figure 18.2. Cross section of the Desert Iguana's vomeronasal system. (*drawing by Frances Zweifel*).

monsters use their vomeronasal system to find food. In separate tests, two investigators dragged a mouse and a pigeon egg along a surface so that chemicals from the food items would remain on the ground. Gila monsters traced the mouse or egg by touching their tongues repeatedly to the pathway. Even when the mouse was in clear view, the Gila monsters still followed the chemical trail. Gila monsters exhume and eat the eggs of other reptiles; tongue extrusions may help locate them.

　　While some lizards use the vomeronasal system to find food, many seem to use this system for different reasons. For example, my studies of the behavioral ecology of Yarrow's spiny lizard, most of which were done at the American Museum of Natural History's Southwestern Research Station in Arizona, show that this animal does not use its tongue for food seeking; only one of more than 2800 tongue extrusions immediately preceded feeding. This lizard is primarily a sit-and-wait predator, as opposed to an active searcher. It perches on a favorite rock for hours, darting out to capture any food that appears, watching for predators, chasing away unwelcome members of the same species, and watching for mates during the breeding season. The lizard uses its tongue for other purposes as well as for feeding and drinking. I once thought that perhaps tiny insects were being captured as the tongue touched the rocks, but a careful search of the rocks, plus an examination of the stomach contents of several lizards, convinced me otherwise. I also searched carefully for small drops of water and found the rocks to be perfectly dry.

Since many lizard species frequently stick out their tongues when they are not looking for food, we can assume that some other type of information is being obtained. In particular, the tongue may help some lizards in sex or species recognition. Male western banded geckos, for example, respond to a chemical stimulus emanating from the skin of the female. This stimulus may serve to attract males to females and to help differentiate males from females. One investigator exposed courting male banded geckos to partly anesthetized males and females that had their tails surgically exchanged. Males often touch their tongues to various parts of the female body, and normal courtship involves biting and holding the tail of the female. Courting males in this experiment rarely gripped male parts. In several cases, males gripped and held the female tail on male individuals. Other male lizards, such as the side-blotched lizard and the large chuckwalla, both found in the western United States, also touch their tongues to females during courtship. The side-blotched lizard often samples the flank and groin areas while the chuckwallas touch many areas of the female body.

Maternal behavior may also be aided by the vomeronasal system. Few reptiles exhibit any sort of maternal behavior, although there are some obvious exceptions such as female alligators, which defend their nests as well as the hatchlings, and pythons, which incubate their eggs by coiling around them. A few lizards also exhibit minimal maternal care, and such care may sometimes be aided by information brought in by the tongue. Broad-headed skinks often lick and turn their eggs and may even spend time brooding them. In one experiment, when eggs of different kinds of lizards were substituted for skink eggs, females refused to incubate many of them; varnished and waxed skink eggs were ignored as well. Since the investigators could not rule out the role of smell in this experiment, they suggested that broad-headed skinks may use some combination of smell and the vomeronasal system to recognize their own eggs.

For some lizards, recognition by males through tongue flicking of chemicals produced by females may help ensure that mating occurs between members of the same species. The vomeronasal system helps adult lizards recognize others of their species for nonmating purposes too. For example, newly hatched green iguanas extrude their tongues at the nest opening prior to complete emergence. Once outside, juveniles proceed slowly while touching their tongues to the ground and other hatchlings. At this age these animals are gregarious and investigators believe that detection of telltale chemicals through tongue flicking may help the young iguanas stick together.

The chuckwallas in California may actually mark the substrate with secretions. These markings could then indicate to a lizard that another member of the same species lived there. Such markings would also help to delineate territorial boundaries. Snakes that prey on lizards might be detected in a similar manner. If snakes are depositing chemicals so that members of a single species can find or avoid one another, lizards may have evolved the ability to chemically detect the presence of snakes.

Thus the vomeronasal system has been implicated in food seeking, sex recognition, courtship, species identification, orientation, maternal behavior, territorial behavior, and predator detection. Many of these observations and studies are anecdotal, however, and need to be tested in a quantitative and controlled manner for various species. Since none of the studies had thoroughly investigated the reasons for tongue extrusions by a single species, my research team decided to examine in greater depth the role of the vomeronasal system in Yarrow's spiny lizard. Karen Gravelle and Barbara Bissinger, then doctoral candidates at the City University of New York, worked closely with me on this project.

We began by determining that the mean number of tongue extrusions for males and females, be they juveniles or adults, is approximately ten an hour. Under certain circumstances, however, tongues are extruded far more frequently. For example, these lizards touched their tongues to the substrate (rocks, logs, trees, or the ground) an average of 26 times in the hour following morning emergence, and lizards removed from their home ranges extruded their tongues 31 times each hour. This increase of two and one-half to three times the normal rate of extrusions indicates that Yarrow's spiny lizard uses its tongue to monitor certain chemical changes in the environment. We suggest that these animals are monitoring their chemical environment most carefully when they first emerge in the morning and when exploring new areas.

But what types of chemicals are being monitored? As suggested earlier with chuckwallas, there is a distinct possibility that lizards are monitoring chemicals that reveal the presence of other members of the same species. Pheromones deposited by conspecific lizards may help to mark territories and/or assist in finding individuals for mating. Laboratory experiments showed us that the incidence of tongue flicking increases significantly when Yarrow's spiny lizards are placed in empty cages where other members of the species once lived. This increase, which is found in newly born lizards as well as adults, suggests that the lizards find something interesting on the substrate, presumably pheromones deposited by other lizards, and continue to explore this information with their tongues. Yarrow's spiny lizards are territorial but do not constantly guard their territories. A chemical marking system could identify well-established territories for a newcomer even when no visual cues are available.

By focusing on the relationship of tongue extrusions to mating, Karen Gravelle has obtained good evidence that pheromones exist. In her experiments, four adjoining outdoor pens were constructed. Three pens housed one lizard each: an adult male, an adult female, and a juvenile male. The fourth was left empty, as a control. The residents were taken out during testing periods, but remained in their home pens at all other times. Adjoining pen walls were removed to make one large pen during the 30-minute testing periods, and fifteen adult males, one at a time, were placed in the center of the pen. The incidence of tongue extrusions increased when the males entered the area where the adult female had resided. One male stopped at the female's favorite perch site and thoroughly explored this area with his tongue. These observations provide evidence that males chemically detect the presence of females during the mating season. Since individual Yarrow's spiny lizards are not continually active, a male that discovers a fresh chemical deposit from a sexually receptive female might remain in the area until she emerges again. Once the female emerges, vision becomes the dominant sense involved in the mating process.

Although Yarrow's spiny lizards can chemically detect other members of their species, the source or sources of the pheromones are still uncertain. Several possibilities exist. Many lizards, Yarrow's spiny lizard included, often rub their chins on the substrate. They may have itchy chins or they may be depositing a pheromone. Histological studies have not revealed any external glands along the chin of Yarrow's spiny lizard, but the mouth, just inside the lip line, is loaded with glands. Substances from these glands may exude from the mouth and then be rubbed by the chin onto the substrate. Interestingly, the males in Gravelle's study exhibited more defecations and pelvic rubs in the female's home pen than in any of the other three pens. These could be ways to mark an area. Many lizards have large pores along the inside of the femur, and these may secrete substances that can be deposited by pelvic rubbing. In addition, many glands exist in the vicinity of the cloaca (the single reproductive and excretory opening), and pelvic rubbing can smear glandular secretions, as well as urine or feces, onto the substrate. David Duvall of the University of Wyoming found that cloacal exudates of both male and female western fence lizards (which are closely related to Yarrow's spiny lizard) elicit tongue extrusions. The lizards may not even have to make an effort to mark an area; resting the cloacal or femoral pores on the substrate may automatically result in the application of chemicals.

One thing that particularly intrigued us was the low rate of tongue extrusions by Yarrow's spiny lizard compared with that of other species. Why do some flick out their tongues more than others? Barbara Bissinger initiated a study in which we examined rates of tongue extrusions for fourteen species representing six taxonomic families of lizards. All were observed under similar laboratory conditions, some at the Bronx Zoo, some at the American Museum of Natural History, and some at our City College laboratory. In a laboratory setting, Yarrow's spiny lizard, a representative of the family Iguanidae, extrudes its tongue approximately 16 times each hour—not statistically different from the mean rate of 10 times per hour that we observed in the field. Both values represent relatively infrequent use of the vomeronasal system. A second iguanid, the neotropical tree lizard *Enyaliosaurus clarki,* extrudes its tongue approximately 26 times each hour. Contrast these relatively low rates of tongue extrusions with those of two Australian members of the family Scincidae, the common blue-tongued skink and the blotched blue-tongued skink. The common blue-tongued skink extrudes its tongue an average of 257 times each hour; the blotched blue-tongued skink, 330 times each hour. If these rates seem extraordinary, consider these three members of the family Teiidae: the western whiptail flicks out its tongue an average of 456 times each hour; the checkered whiptail, 623 times per hour; and the Chihuahua whiptail, more than 700 times an hour.

Flicking out the tongue hundreds of times each hour takes considerable energy and probably results in some water loss through evaporation. This behavior must therefore serve some purpose. Whiptail lizards must rely heavily upon the information brought in by the tongue for many aspects of daily life. Smell also seems to be important for these lizards, which often sniff in a manner similar to that of humans. At the other extreme, iguanids such as Yarrow's spiny lizard rely much less on information gathered by the tongue. The skinks, showing an intermediate incidence of tongue extrusions, form an intermediate group.

Most iguanid lizards are colorful, highly visible, and exhibit obvious behavioral displays. For example, territorial behavior for the dark Yarrow's spiny lizard involves slashing its blue tail from side to side, exposing bright blue sides by flattening its body, presenting these sides to an intruder, hopping sideways, doing pushups, and extending a bright blue gular area under the chin. These animals spend a great deal of each day basking and waiting for food to come to them, and all these activities make them highly visible to other iguanids. The Jacobson's organ is reduced, but functional, and the tongue is broad and fleshy. The function of the vomeronasal system is important, but vision is the dominant sensory system.

Skinks are less visually oriented and show various stages of reduction in eye size, as compared with other lizards. Skinks are relatively secretive and hard to find. Rather than spending the day sitting on visible perches waiting for prey, they actively search for food, often under cover. Basking is infrequent. They seem to rely more upon chemical signals to find or avoid one another and to obtain food. Anatomical evidence supports the hypothesis that vomeronasal chemoreception is of greater importance for scincids than iguanids. The skinks' Jacobson's organs are well developed and may be aided by the slight bifurcation of the tip of the tongue. A long, thin bifurcated tongue is thought to deliver chemicals to the Jacobson's organs more efficiently than a thick, fleshy tongue.

The tongues of whiptails, which are extruded more often than those of the other lizards, are exceptionally long, thin, and forked, resembling those of snakes. The Jacobson's organs are well developed in these swift-moving terrestrial lizards. Visual communication among whiptails is not obvious and chemical communication may aid in social organization. These lizards do not seem to communicate with head bobs, pushups, tail slashes, and other visual signs typical of iguanid lizards.

Other taxonomic families showed varying rates of tongue extrusions that also correlated nicely with the importance of vision and the development of the tongue and Jacobson's organs. Often, members of a single family show similar rates of tongue extrusions. This is not surprising since closely related species share similar evolutionary histories, resulting in many similar factors such as behavior and anatomy. But evolutionary divergences have also occurred within families, and members sometimes exhibit wide variations in the incidence of tongue flicking. For example, we compared field rates of tongue extrusions for Yarrow's spiny lizard with those of the Saint Vincent tree anole, which lives on Saint Vincent in the Caribbean. While Yarrow's spiny lizard extrudes the tongue an average of ten times per hour in the field, the Saint Vincent tree anoles almost never extrude their tongues. Both lizards are iguanids, but in this case they reflect some different evolutionary selection pressures. Although both are highly visual, the anole is arboreal while the spiny lizard is primarily terrestrial. One researcher has suggested that an arboreal form of life requires extremely sharp vision, resulting in massive orbital development; this, in turn, could leave less space for the olfactory and Jacobson's organs. Since these chemical senses are probably of less use in a nonterrestrial existence, presumably the net result was a displaced and reduced olfactory chamber. Although these speculations are untested, terrestrial lizards generally have much more developed olfactory chambers and Jacobson's organs than arboreal lizards.

We have pieced together a great deal of interesting information about the lizard vomeronasal system, but much of it is speculative and a great deal of work remains to be done. Lizards are complex animals that rely upon a variety of sensory inputs. Even though a lizard may have excellent vision, it can rely upon other sensory modalities, such as chemoreception, to aid in various aspects of its life. Not all lizard species rely on tongue extrusion to the same degree. But for many lizards, tongue flicking serves a variety of useful ends, helping them make their way in the world.

19

Peaceable Peccaries

JOHN A. BYERS

In 1633, an anatomical account of a New World mammal was published in the *Philosophical Transactions* of the Royal Society of London. The beast had been brought alive to England and when it died was given to Dr. Edward Tyson for examination and dissection. Tyson had in hand a male of a species known today as the collared peccary, *Tayassu tajacu.* His wonderfully descriptive name for the animal was the Mexico musk-hog. Spanish explorers of the time also knew the animal and referred to it as the javelina (from the Spanish *jabalina,* or "spear," a reference to the animal's large canine teeth).

The collared peccary is one of three extant species in the family Tayassuidae. The tayassuids diverged from the true pigs, or Suidae, in the early Oligocene, about 40 million years ago, and differ from true pigs in several important respects. Like pigs, however, peccaries possess a tough, disk-shaped snout and have a proclivity to root. Although they were once cosmopolitan, peccaries are confined today to the New World. The collared peccary occurs over a huge range between Argentina and the American Southwest; its close relatives, the white-lipped peccary and the chacoan peccary, are found only in the tropics and subtropics. In the United States, collared peccaries occur in the southern regions of Texas, New Mexico, and Arizona.

Peccaries and pigs differ in many ways. To begin with, the former are much smaller: collared peccaries are about three feet in length, stand about a foot and a half at the shoulder, and weigh about fifty pounds; wild boars (from which domestic pigs were derived) may weigh up to ten times as much. The coats of peccaries are composed entirely of large, coarse bristles; those of pigs vary, but in general they have hair as well as bristles, the latter being more scattered, shorter, and less robust than the bristles of peccaries. Pigs have a simple stomach; peccaries have a multichambered stomach resembling that of the ruminants in form and, somewhat, in function. Pigs have rather primitive feet; peccaries have more specialized feet that show adaptations for running. The canine teeth of pigs curve outward, as in the wart hog; those of peccaries are straight and interlocking. Finally, the two groups differ with respect to litter size and social organization. Pigs give birth to large litters of small, relatively helpless young; mothers spend much time alone caring for their young in a nest or den; and there are no permanent social groups. Peccaries do have permanent social groups and females bear litters of only two young, which are able to walk soon after birth. As a result, mothers are able to rejoin their herd quickly.

I studied peccaries on the 3-Bar Wildlife Area, a sixty-square-mile research preserve maintained by the Arizona Department of Game and Fish. I was drawn to the collared peccary because, in the literature, there were tantalizing intimations that the species was highly social. A major goal of my earlier research had been to discover whether specific types of early experience lead to specific types of adult behavior in animals. More generally, what developmental strategies does each species use to achieve its level of social organization? This is, of course, a huge question, and one way to begin attacking it is to look for broad patterns at the ends of the spectrum of social organization—that is, in highly social and in nonsocial species. My intimations that peccaries would be at the high end of the spectrum turned out to be correct.

The social unit of peccaries is a cohesive herd, composed of equal numbers of males and females. This reflects a sex ratio at birth of 1:1 and the fact that individuals rarely disperse into new areas. Herd size in the American Southwest, and throughout Central America and South America, in those areas where reliable counts have been made, averages twelve to fifteen members. There is essentially no sexual dimorphism. Within herds, animals remain close to each other, and individuals become visibly upset and may utter loud, squalling cries if they lose contact with their fellows. Herds therefore act as a unit in which all individuals move, feed, and rest together.

Each herd occupies an exclusive home range about one square mile in size. Within each home range a network of peccary trails connects feeding locations and traditional resting sites, or bed-grounds. The animals give the strong impression that they possess a detailed "cognitive map" of the home range; when undisturbed, their movements are deliberate, and they choose energetically efficient paths (by angling up a hill rather than attacking it head-on or by contouring around the head of a canyon rather than crossing directly by going down and then up). Obviously, the animals know, in some sense, their destination (whether a bed-ground or a feeding ground) and they take the most efficient, but usually not the most direct, route to get there.

Herds also seem to know when they are near the borders of their home ranges. At these times, the peccaries increase the rate of deposition or secretion from their large dorsal gland, a scent organ present in both sexes that lies on the midline of the back, just anterior to the tail. Tyson made much of the dorsal gland in his 1683 article and, after examining in a withering fashion the hypotheses of his contemporaries that the structure was a second navel, a mammary gland, a urinary orifice, an intestinal orifice, or a spiracle for breathing, he correctly identified it as a scent gland:

"But what is most particular in our *Hog*, and makes the greatest wonder; and differences it, from any other Animal I know of in the World; is the *Teat* or *Navill* or *Foramen* rather on the hinder part of the back. All who mention this *Animal*; look on this, as a thing so extraordinary, and so uncommon; that I know not how their amazement has so far clouded their reason, as to betray them into most extravagant Conjectures, and opinions concerning it."

Scent marking is probably the way in which adjacent herds define the boundary between their home ranges. This seems plausible because the rate of actual interactions between herds is extremely low; in a year of field observations I saw herds come together only twice. In one of these instances, there was a flurry of short chases back and forth, then both herds drifted apart and bedded down within 1000 feet of each other. In the other, the same two herds were together, feeding peacefully along the border between their home ranges when I found them; they slowly fed and ambled in separate directions.

I tested the idea that scent marking might be important in boundary definition by erecting fence posts at quarter-mile intervals along a ridge where the home ranges of three herds met. Peccaries investigated the posts but did not mark them. I then collected dorsal gland secretions from a captured member of a herd five miles away and placed it on the posts. Residents responded to this new odor by marking on top of it, often with such vigor that I could smell the posts from some distance and could see dorsal gland secretion dripping from them. This result suggested that peccaries recognized an unfamiliar dorsal gland scent and re-marked it to indicate that the area was occupied.

The dorsal gland, besides being used in scent marking, figures prominently in a cooperative social action I call the "mutual rub": two animals approach each other and stand side by side so that each rests its head against the other's rump; each then vigorously rubs the side of its head up and down over the other's rump and dorsal gland. This activity lasts about five seconds, and no doubt smears the oily secretion of each partner's dorsal gland onto the head and back of the other. Peccaries also spend much time nuzzling each other and lying in close contact—two activities that further promote the transfer of the secretion. The overall result is that each animal becomes an effective broadcaster of the dorsal gland scent, so much so that when herds were directly upwind from me, I could sometimes detect them by smell. This suggested to me that peccaries, with their keen sense of smell, use a "cloud" of dorsal gland scent that surrounds each herd as a cue to judge their distance from the herd's center. Vision is certainly less important. Peccaries have notoriously poor eyesight and, if the wind is right, will blunder to within a few feet of an unconcealed, but motionless, person before taking flight.

Hearing may be the most important orienting mechanism. On several occasions I was very close to feeding herds as they drifted past me, and each time I was impressed by the sounds that envelop a feeding herd: the click of hoofs on rocks, the crunching and smacking sounds of chewing, and the periodic low grunting all herd members make. The low grunt is a repetitive flat sound, similar to the grunts made by pigs. Lost peccaries emit the sound almost constantly and appear to listen for a return call; they quickly approach animals in the vicinity that respond. Thus, group cohesion seems to be promoted primarily through hearing and olfaction.

Why do collared peccaries stay close together and why is vision not an important means of doing so? The species probably evolved from other peccaries in the Pliocene rain forests of South America. Predation by large cats is today, and no doubt was then, a serious threat, constituting a selective pressure strong enough to cause individual peccaries to bunch together. In the dense rain forest understory, these low-slung beasts would have had trouble using vision to keep in contact with each other, but they could have used sounds and smells (to which vegetation is more "transparent") to do so. Bird watchers are familiar with this phenomenon. In thick vegetation, far more birds can be heard than seen.

The tendency for peccaries to stay close together is so strong that most individuals probably never leave the herd into which they are born. As a result, herds most likely comprise a group of closely related animals. This, in turn, has probably been responsible for the evolution of the highly cooperative behavior that herd members extend toward each other. (Chacoan and white-lipped peccaries also live in herds, and white-lipped groups can exceed 200 individuals. Essentially nothing is known about the constancy of herd composition or social behavior in these two species.)

Cooperative behavior among peccaries can be crucial in winter. In central Arizona, winter nights are usually chilly and sometimes downright cold. On one memorable morning at the 3-Bar, the temperature before the sun rose was 17°F (-8.3°C), and there was a light covering of snow that capped the saguaros and made them seem to shiver. Peccaries are typical tropical mammals. Because they have no heat-retaining wooly undercoat, peccaries quickly begin to expend extra energy to maintain a constant body temperature when they are exposed to cold. As a result, in the winter they are diurnal, moving and feeding during the day. Huddling together on winter nights is essential if the animals are to stay alive. (As summer nears, peccaries become increasingly nocturnal, moving and feeding at night and resting in a shaded bed-ground throughout the day.)

One typical winter morning, I arrived at a vantage point above a herd's bed-ground before sunrise and before the animals were up. Rustling amid the turbinella oaks began as the sun appeared, and soon stretching, yawning, and shaking animals drifted into view. There was a concentrated bout of mutual rubbing between herd members, then the animals began to move in a rough single file from the bed-ground. They traveled only a short distance until, reaching a sunny spot, they began to feed.

During the winter much of the highly nutritious food on which peccaries usually depend, such as acorns, bulbs, mesquite beans, and cactus fruit, is unavailable, and for several months the animals may subsist almost entirely on prickly pear cactus. Lyle Sowls, leader of the Arizona Cooperative Wildlife Research Unit, has shown that captive peccaries can survive for several months on prickly pear alone but that they eventually deteriorate on such a diet. The plant is important for the animals because it sustains them through the lean winter period. The pads, or cladophylls, of the prickly pear cactus, with their long spikes and tufts of spines, do not seem to be inviting food items, yet the peccaries, with apparent disregard for the spines, attack them with impunity. Occasionally, individuals will worry a pod loose from the plant, then paw at it as if to break off some spines. More frequently, they simply approach a plant and bite directly into it.

In the herd I was observing, feeding animals remained close together and often several individuals fed side by side from the same plant. Occasionally, squabbles broke out when one animal approached and began to feed next to another, but more frequently the animals simply fed together peaceably. In many instances, two animals were able to approach and feed so close to each other that their snouts almost touched. Juveniles, in particular, were able to approach adults and feed mouth to mouth with them. Juveniles even snatched food dangling out of the mouth of an adult without suffering retribution.

After a bout of intensive feeding, the herd began to travel rapidly in a particular direction, with animals feeding in a more choosy and desultory fashion. Moving and feeding continued until between 10:30 and 11:00 A.M., by which time the herd had traveled approximately a half mile from where it started and was nearing another bed-ground. As is often the case, this site was surrounded by a radiating network of trails, and when the animals encountered a trail, they quickly moved onto it and walked steadily toward the bed-ground.

As the animals entered the bed-ground a striking event took place: all herd members began to play in a wild and comic manner. Such play bouts are frequently started by juveniles, but most adults participate in them. Mutual chasing around a bush and one peccary flopping on the ground to mock bite with a partner standing over it were particularly common. My favorite play action involved one animal running full-tilt downhill toward another lying flopped on its back; the two would clamp jaws, and the running peccary's momentum would carry it sailing over the other, rotating in a slow arc, like a diver in layout position, until it crashed heavily on its side. As is true of their social behavior in general, peccaries show great behavioral diversity while playing. Play bouts lasted about fifteen minutes, and juveniles tended to play longer than adults. Eventually, all animals reclined close together, many lying side by side.

The herd rested for several hours. Individuals got up and fed briefly or changed places to recline next to another herd member or moved in and out of the sun to keep comfortable. Much quiet, amicable behavior, including sniffling, nuzzling, and lying in close contact, could be observed during this time. When the animals began to move again, they did so rather suddenly. More feeding followed, with the animals moving rapidly. As darkness fell, the herd was close to the bed-ground where it would spend the night.

A day in the life of an individual peccary is thus dominated by feeding, slow movement, and resting, with other animals always close at hand. The day is punctuated by brief feeding squabbles, mutual rubs, and most conspicuously by play. The single most revealing and important fact about peccary social organization is that the average distance between animals is about ten feet, or three body lengths. Individuals are born into a herd and are literally surrounded by familiar animals throughout their lives. By this criterion alone, peccaries are more social than most other ungulates. Moreover, as I have intimated, they also display a level of cooperation that makes them unique among the hoofed mammals.

Cooperation takes several forms. First, there is some degree of cooperative feeding, or food sharing. This practice may occur between adults, but it is especially prevalent between adults and juveniles. All juveniles are apparently given carte blanche by all adults, and juveniles are not in the least hesitant to use their special status, often aggressively driving an adult from a food source, as well as snatching food from an adult's mouth. Adults could easily dominate juveniles in these encounters, but they do not. This social convention is probably important for juveniles in several ways: it provides for mostly positive social interactions between juveniles and their future peers; it gives juveniles access to high-quality foods; and it probably teaches them what types of plants should be eaten. All adults extend these favors toward *all* young; non-parents are as tolerant as parents.

Feeding interactions illustrate a more general phenomenon: adults are tolerant of juveniles in all respects. In bed-grounds, juveniles clamber over the reclined herd and try to wedge themselves between adults; on trails, they are often able to shoulder adults aside. In one typical instance, two small infants stood beneath an adult's chin and repeatedly snapped at its mouth as it tried to feed. The adult was forced to elevate its snout in order to chew, and in this position it began to walk forward. The juveniles darted between its front legs and caused it to stumble. There was no retribution.

Adults also act cooperatively to protect juveniles from predators. The canine teeth of peccaries in both sexes are long, robust, and self-sharpening since the back of the bottom canines slide against the front faces of the teeth on top. Adult peccaries do not tolerate coyotes or bobcats. They quickly drive them away, while often ushering juveniles in another direction. The response of herds to humans, and probably to large predators such as bears or mountain lions, is different. Here there is no attempt at defense; the animals simply run away. If the disturbance is mild enough (as when the herd picks up the scent of a distant human observer), there is a well-organized retreat, with one or two adults in the lead, followed in single file by the juveniles, then the rest of the adults. If the disturbance is more immediate (as when the observer blunders into the midst of a herd), the animals tend to scatter in all directions in an every-peccary-for-itself manner. This set of strategies makes sense. If defense or organized retreat is possible, it is carried out. If the danger is so large and imminent that defense would be futile, the herd's scattering may tend to confuse a predator. It also would tend to draw attention away from the helpless juveniles, which in these situations seek some kind of cover and remain motionless. Herds quickly regroup after a disturbance, individuals using the low grunt to locate one another.

A third form of mutual assistance is cooperative nursing. Unlike females in most ungulate species, peccary females will allow young other than their own to nurse. Juveniles approach from the rear and suckle while the female stands. Females do not turn around to check the identity of these young, and it seems that most young are able to nurse from most females that have milk. On several occasions in herds that had many new young, there was a snapping, jostling horde of hungry young at the rear end of one female. There were obvious injustices—with larger young pushing small ones aside—but the nursing females continued to munch cactus complacently, as if the storms behind them did not exist. Milk production in mammals is an energetically expensive business, and females of most species usually make sure that they give milk only to their own young. In many ungulates, females will not tolerate an alien juvenile in their vicinity while they are nursing. I thus found it striking that peccary females seem to distribute milk indiscriminately.

Peccaries are also unusual mammals in that males do not compete actively for mating rights. When a female is in estrus, there may be slight increase in aggressive behavior, but more often than not, pairs court and mate with other males quite close by. On one occasion I saw a female mate with one male, then turn and court and mate with another while the first stood idly by. I found it curious that males seem to give up individual reproductive success. The most likely explanation is that the males in a herd are so closely related to each other that for any individual the difference in real reproductive success between a female producing his offspring, or the offspring of another male is so slight that it is not worth fighting about. It is not known if in-breeding reveals deleterious recessive genes in peccaries or if there are social mechanisms that preclude matings between very closely related individuals.

The low level of competition between males is also reflected by the lack of sexual dimorphism in peccaries. Males and females are equal in body size, and both sexes have large, sharp canine teeth. In other species that have mating systems in which males fight for access to females, males are usually much larger than females and tend to have large, conspicuous fighting and display structures, such as the horns and antlers in many other ungulates. The lack of sexual dimorphism in peccaries suggests to me that the species has had a substantial evolutionary history of reduced male-male competition.

The collared peccary is thus, by many measures, a highly social beast. The animals in a herd are cohesive and possess two specialized ways of maintaining close contact (the mutual rub and the low grunt). There is cooperative feeding, mutual tolerance of young by all adults, regular play that involves all herd members, cooperative nursing, and reduced male-male competition. The species is close to one extreme of the spectrum of mammalian sociality and was indeed appropriate for my study of behavioral development.

The development of behavior and the nature of early experience in peccaries is another story in itself. The gist of what I found is that in development, as in adult social behavior, peccaries are different from most other mammals. In other highly social species, juveniles interact amicably with each other and in this way probably form the bonds that draw them together as adults. Peccary juveniles barely interact with each other for the first seven months of life and when they do interact, it is likely to be in a fight over access to food or a female's teat. Yet these juveniles eventually grow up to become amicable, cohesive, and cooperative adults. Adult tolerance of juveniles is probably vital in this species, and juveniles seem to be brought into the social group via their interactions with adults, not their interactions with each other. There are intricacies in this process that I will try to sort out in future field seasons.

20

New Theory on a Fabled Exodus

KAI CURRY-LINDAHL

"Lemming migration," "Multitudes of lemmings," "Wells at X-ness full of lem-mings," these were some of the headlines that caught the eye of the Swedish newspaper reader at regular intervals during 1960 and 1961.

What were the real facts behind these headlines? Was Lapland truly in-undated with lemmings? Was there a dense carpet of these small rodents pressing irresistibly from the mountains toward mass suicide in the sea?

For centuries, stories about the Norway lemming *(Lemmus lemmus)* have spread all over the world. In 1532, Ziegler of Strasbourg published a treatise on the lemming, based on information obtained in Rome from two bishops from Nidaros in Norway. He related that in stormy weather lemmings fell from the sky in enormous numbers, that their bite was venomous, and that they perished by thousands when the grass began to grow in the spring. In the chronicles of the march of King Charles XII's soldiers over the mountains between Jamtland and Norway in August 1718, Jöran Norberg wrote, "People maintain that the clouds passing over the mountains leave behind them a vermin called mountain mice or *lemmings* by the inhabitants; in size they are as big as a fist, and they are furry, like the guinea-pig, and poisonous." The same legend of the cosmic origin of the lemming is also found among the Eskimos, whose name for one Alaskan species means "the creature from space."

The Norway lemming is found in all parts of the Scandinavian high mountain range, as well as in the highlands of northeastern Swedish Lapland and in Finland. Lemming population rises periodically, some years to enormous numbers, in all the vegetation belts of the mountains; in the intervening periods it is very low.

In most parts of the mountains, the spring migration of lemmings during a "normal year" (not an "eruption year") need not cover great distances toward lower altitudes; winter and summer quarters are generally close to each other, usually in the same belt of vegetation, most commonly above the willow region. Thus, above an altitude of 2500 to 3300 feet, migration is generally horizontal.

In summer, lemmings seek shelter in natural depressions and cavities in the ground, or make tunnels in the ground vegetation. Such refuges, which are often no bigger than the animal itself, are used regularly, and depending on the condition of the ground, the lemmings' paths can be discerned in the carpet of lichen.

From May to August, the animals prefer to inhabit moist, stony ground partly covered by sedges (*Carex*), willow shrubs (*Salix*), and/or dwarf birch (*Betula nana*). Such habitats provide the animals with food and the hiding places necessary for survival in an area of many predators—the stoat, or ermine, the weasel, the rough-legged buzzard, the common raven, the long-tailed skua, crow tribe, and snowy owl. Water does not seem to inconvenience the lemming during summer, for its fur is water repellent, and the animal behaves almost as if it were aquatic. However, it needs dry holes for reproduction, as the newborn young are sensitive to moisture and cold.

The situation is quite different in winter. In autumn a move starts for drier places, because in the winter damp ground combined with severe cold can mean death for the adults as well. The lemmings can usually live in comparative safety under the snow, which protects them from cold and from enemies. If rain and frost should blanket the vegetation with an ice sheet before the snow cover is established, however, the result for the animal may be fatal, as food-gathering then becomes a serious problem.

Although many summer predators are completely absent in winter, the stoat and weasel hunt under the snow. They are the lemmings' only important enemies in wintertime, and even these are relatively rare in the lichen region during winter. Thus, the lemming population above timber line is practically free from predation until early spring, when the lemmings first venture above the snow. Beneath the snow, lemmings construct extensive passages and build round nests of grass that are sometimes attached to willow shrubs, and may be seen hanging on twigs after the snow melts.

During an ordinary winter, without unusual vacillations in climatic conditions, the rodents take advantage of their protected situation to breed. The lemming is so prolific in some years that, in view of our experiences in 1960 and 1961, which I will go into, I am inclined to suspect that winter breeding is a necessary condition for a subsequent population among rodents, in the light of present knowledge, although we have found some indications that the field vole (*Microtus agrestis*) may also breed in winter. In Scandinavia, the lemming may have survived in some refuge areas during final Ice Age glaciations. If the animal entered the

Ice Age much as it is now, then its ability to breed under the cover of a snow blanket would obviously have favored it. On the other hand, its specialization may have developed only after the onset of the Ice Age in its habitat.

For centuries people have speculated on the reasons for the rapid increase of lemmings. The explanation is probably that the species, hidden from view from autumn to spring, is able to build up several generations in a single season. But why does the lemming manifest such marked population increases in certain years? In addition to a high breeding potential, certain environmental factors are inherent in a population explosion:

1. Generally favorable climatic conditions, which in turn affect both the supply of food and the ability to take advantage of it during a great part of the year, and permit survival of the young in their passive period in the nest. Early springs and late autumns would therefore by propitious.

2. Proper climate during the winter periods when newborn young are in the nest. Mild weather and thaws may then be fatal, and it is possible that severe cold has the same effect.

3. The almost complete absence of predation pressure in winter.

But if the long winter and the blanket of snow are so advantageous, as claimed above, how, then, can early springs and late autumns also be favorable? The fact is that the supply of food beneath the snow is limited, so that the idyll there may become a fatal trap if it lasts too long, particularly during the years when lemmings are most numerous. It is very likely that this trap snaps rather often—one explanation as to why population explosions occur only infrequently.

As mentioned above, the seasonal migrations of lemmings from one relatively nearby biotope to another are not the same as the dramatic marches that have made lemmings so famous. In light of the 1960–61 investigations, when there was an eruption of Norway lemmings, it seems that the populations normally move within their natural biotopes—the extensive mountain heaths in the lichen belt, downward to willow, birch, and conifer regions, from whence they disperse overland.

These less noticeable vertical movements of lemmings flow, as it were, slowly down the mountain slopes. Individually the movements seem to be at random—a straying in different directions, both high in the mountains and lower down. Breeding continues during these movements, but females that are in an advanced stage of pregnancy settle at whatever level they may be, even when far away from the lemmings' most favorable and densely populated habitats, which lie high up on the mountain heaths.

It is probable that the lemmings were on the move throughout practically all the summer months of 1960, but their movements were individual and scattered, which made it impossible to determine direction, speed, and the number of animals taking part. It was clear, however, that three waves of lemmings flowed downward from the upper vegetation belt into the conifer forest. The first was at the end of May, when the animals began to appear or to increase greatly in numbers in the upper coniferous tracts on both sides of the mountains that lie between Sweden and Norway.

A new wave became noticeable about a month later. This rise in the population was clearly due both to on-the-spot breeding and to the arrival of lemmings from higher altitudes. Finally, a third wave broke over the conifer belt at the end of August and September, in some places as late as October. In the last case, too, the frequency peak was probably combined with breeding activities.

The mass migrations, on the other hand, seem to occur only in certain topographical situations—if a long lake stops the rodents' slow, almost invisible progress, or two rivers meet and the lemmings are, so to say, caught in a funnel. In these and similar situations, there is a continuous accumulation of animals. Finally, the concentration is so great that a certain panic reaction results—a kind of mass suggestion. This is expressed in a reckless march that need not follow any special direction, but may go north, south, east, or west, uphill or downhill, over rivers and lakes, and sometimes to the sea—particularly in Norway, where the mountains are close to the shore. The march may also go from the mountain heaths up over the glaciers. Thus, a panic-stricken flight in the form of a mass migration may begin up on the mountain heaths, but there, too, overpopulation seems to be the main cause. There is much evidence to suggest that the eliciting factors of mass migration are a kind of psychosis, possibly owing to the competition with other individuals for sheltering holes and territory, but not, as far as is known, for food; even when lemmings are most numerous, food is available in the vicinity.

It is quite possible, however, that the food of the Norway lemming is far more specialized than has hitherto been believed, and if this is so, lack of suitable food may be of decisive importance. Diseases also have been advanced as one explanation for the mass migration phenomenon, but migrating populations have not been too ill to settle again and breed in new territories. In addition, many animals remain in their original habitat. The idea that lemming migrations always end with death is therefore erroneous, as is the fanciful belief that they form a dense carpet of moving bodies during mass movements.

In the New World, it has been found that during a peak year vegetation has been completely devoured locally, and this is believed to have caused mass movements. But migration of the North American species—the brown lemming *(Lemmus trimucronatus)* and the collared lemming *(Dicrostonyx groenlandicus)*—is unusual and irregular, and cannot be compared to the periodic migrations in the Fennoscandia region. While in the latter area the movements of lemmings and their dispersion over great areas reduce the risk of food shortages, it is not yet known whether any special plants in the lichen region are of particular significance for the animals' health and well-being. If it should be discovered that certain plants are vital, their decrease during peak population years might well be a prime cause of migration. Perhaps a prolonged lack of a certain vegetable food in the diet contributes to endocrinological disturbances and the unexplained mass deaths to which lemmings are subject.

An interesting situation existed at the end of July, 1960, in the conifer region along the River Graddielva in Norway, west of the mountains of the Pite lappmark area in Sweden. There were swarms of lemmings. They were particularly numerous on a north slope that angles down toward the river, and is close to the junction of two watercourses. Tunnels and nest holes were everywhere in the moss, and there were large, collective heaps of excrement. Well-trodden miniature paths, like small canyons worn in the moss vegetation, formed a dense network. Even during the hottest hours of the day (in a heat wave), with temperatures up to 88° F (31°C), a number of Norway lemmings were active; they were only a minority of the enormous numbers that infested the place during the cooler evening and night. In the daytime most lemmings remained in the holes that were at their disposal, while at night the animals sometimes congregated in great flocks along the bank of the river.

Many individuals swam across, but the majority hesitated. They ran up and down along the shore; some threw themselves into the water but then turned back at once. Obviously, most had not been affected by the innate urge that would probably soon start the living avalanche on its way.

Lemmings swim very fast, with their heads held high. Sometimes they try the current in different places before they finally set out. They often swim diagonally against current until eventually they reach the opposite bank. Occasionally, one jumps up on a stone in the middle of the stream and rests there. During the swims in the Graddielva there was no question of a panic-stricken flight, for the lemmings crossed one by one and almost invariably only after they had carefully chosen a suitable spot for the crossing.

In a spite of this seemingly cautious behavior, it was clear that the population was in a state of constant tension. The restlessness may have occurred because the habitat was not ecologically optimal for an Arctic animal, and because it was overpopulated. It is likely that the process toward an acute psychological tension was developing, but the threshold value that would release mass migration had not yet been passed. As usual, the calm behavior of the pregnant females contrasted with the feverish nervousness of the other lemmings. The former were methodical and purposeful: although surrounded by the chaos of their rushing, jostling, screaming kindred, they always knew where to find their holes, into which they slipped at the first sign of unusual disturbance or danger.

Unfortunately, I could not stay in this region to see the outcome, but a report from a local observer informed me that three days after I left, practically all the lemmings were gone. The majority had crossed the rivers, but a few had been observed moving back in the direction from which all had previously come. If the observations are correct, the last feature is particularly interesting because it is an indication that lemmings do not necessarily move in their initial direction, which is generally downward. They can also return upward.

At this stage there was no sign of any population crash—another phenomenon connected with lemmings. A drastic and sudden decline of the population almost always follows a peak year. If there is no doubt that predation is often important in keeping populations of small rodents in check (as Frank Pitelka and his associates from the University of California in Berkeley have shown to be the case in Alaska), on the whole the population size for such animals seems to be regulated by a complex of factors (including predators), in combination with self-regulating mechanisms. The Norway lemming is a striking example of this. If neither predators nor food plays a decisive role in the fantastic Norway lemming population crashes, which are almost without parallel in the vertebrate world, what is the cause of their wholesale deaths?

In the first place, every Norway lemming must die sometime, and it does so within the four-year cycle that characterizes the periodicity of the species, for the known life span is less than four years. But the majority of Norway lemmings probably do not die of old age. The strange, abnormal features of their behavior during "explosion years" has been mentioned earlier. Such pathological phenomena, with mass death in their wake, may also appear in extremely dense populations of other species of small rodents before a fatal lack of food arises through overcrowding.

Unfortunately, the physiological and microanatomical phenomena, particularly those concerning glandular and metabolic interrelations, have never been examined in detail in *Lemmus lemmus*. Recently, however, a number of significant features in the endocrine organs of the collared lemming have been described by W. B. Quay at the University of California in Berkeley, who has advanced the hypothesis that in warm and/or stressful conditions, a metabolic derangement occurs that is conducive to abnormal deposits of colloidal material in the walls and lumina of small blood vessels of the brain, including the hypothalamus. The condition has been found in feral collared lemmings during two summers at one of the southernmost outposts of the species, and suggests that the described phenomenon may be significant in affecting natural populations.

Quay's conclusions are highly interesting in relation to the fact that manifestations of great physiological imbalance in *Lemmus lemmus* are more often expressed by individuals that occur in summer in lower and warmer areas, to which the species is less adapted and to which they have been forced by population pressure in upper, optimal habitats.

The physiological picture of what happens in collared lemmings living in warm or stressful conditions does not seem to exclude the theory that other physiological factors are also involved. The view that the dysfunction of the pituitary seems to result in an overproduction of the adrenocortical stimulating hormone, which in its turn overstimulates the adrenal function in such a way that the production of corticoid hormones first increases and then reaches a stage of complete exhaustion, does not seem to be antithetical to Quay's results and suggestions. The metabolic derangement found by Quay is probably only a preliminary phase to the physiological collapse that precedes the breakdown of adrenal function, which in its turn causes the death of large numbers of lemmings and explains the sudden crash of the whole population.

It is not known whether the causes of population crashes, simplified and described briefly here, hold good for the Norway lemmings. During the autumn of 1960, however, dead, intact lemmings were often found. There was a pronounced population decline at the end of that year, but the crash did not appear to be associated with either lack of food or predation, but rather with a physiological imbalance.

However, the whole population did not crash in 1960. We had expected a continuous decline during the winter, but in May and June of 1961, to our surprise, we found dense populations in several areas. Apparently the high was locally still in full swing. Then the situation changed drastically, and in the last two weeks of June most lemmings disappeared. The interesting point was that they did not leave their areas by migration, nor were they considerably reduced by predation.

They just gradually vanished on the spot, apparently dying in their shelters and holes, where, after some digging, we found carcasses which, unfortunately, we could not study pathologically.

Further chemical research will undoubtedly produce answers to some of the fascinating and often dramatic behavior of these remarkable rodents.

Part 4
BEHAVIORAL DEVELOPMENT

Fertilized eggs don't eat, drink, or mate. These and all other adaptive, species-typical behaviors emerge gradually in adult animals through an interaction between the organism and its environment. Unraveling this interaction is the job of the developmental animal behaviorist. But exactly what constitutes the environment of a developing individual? Of course there is the physical environment, such as temperature and oxygen, which must be maintained within fairly narrow limits for ontogeny to proceed normally. But many animals also develop in a social setting, so it is equally important to understand how organisms interact with parents, siblings, peers, and any other conspecifics.

As an example of this social interaction, let's consider the developmental process in ants (and other social insects) called temporal polyethism. In *Novomessor albisetosus*, newly eclosed (i.e., from the pupal stage) workers, called "callows," remain in the nest, tending the brood and queen. Only after two months of these indoor activities do workers of *Novomessor* leave the nest and forage for food. And once they begin this outdoor phase of their life, they never again have any sustained contact with the brood. For many years, such changes in behavior were viewed as developing according to a rigid biological "program," tightly linked to anatomical and physiological maturation. But in a recent study conducted in my laboratory, Philip McDonald and I demonstrated that the development of behavior in these ants is modifiable, depending upon the social environment of the workers. When the mature adult ants were experimentally removed, leaving behind the immature callows and a hungry larval brood, the development of these young workers accelerated. Instead of two months, they began foraging outside of the nest in less than three weeks. And in another experiment in which the callow population was removed, the mature adult foragers immediately reentered the nest and reverted to their brood tending stage development.

The five articles that make up this last section all explore the numerous and varied developmental factors that influence social behavior. We begin with Gottlieb's work on the development of the social bond between ducklings and their parents, a phenomenon to which Konrad Lorenz gave the name "imprinting." Although most studies of early experience in precocial birds concentrated on visual communication, Gottlieb demonstrates how auditory cues enhance this form of early learning.

Continuing on the theme of parent-offspring interactions, but moving from birds to baboons, Luft and Altmann illustrate how the duration and intensity of an infant's dependence on its mother is quite variable, depending in part upon the parent's social status in the group. This is followed by an article by Jones and Bush on how behavioral development in desert-dwelling kangaroo rats is influenced by such ecological factors as the availability of vacant mounds and the abundance of seed caches.

Many immature mammals exhibit patterns of behavior which resemble those of adults but which seem to have no immediate benefit. They run, climb, chase each other, and manipulate objects. Such behavior, which is relatively rarely seen in mature adults, is usually called "play." In an article based on studies of Hooker's sea lions, Bruemmer describes the birth and early behavior of sea lion pups and shows how detrimental their play activities can be when exploring pups get in the path of territorial bulls. But on the other side of the coin, in an article about play in a variety of nonhuman primates, Dolhinow shows the importance of play activities in providing necessary practice for increasing the efficiency of adult behavior.

21

Components of Recognition in Ducklings

GILBERT GOTTLIEB

During the past fourteen years, well over a hundred research articles have reported laboratory studies of the "following-response" and "imprinting" in young precocial (nidifugous) birds. Such animals (ducks, geese, swans, chickens, turkeys, quail) can locomote shortly after they are born, and it has been found that, independent of feeding or other conventional "rewards," they quickly attach themselves to the first moving (or otherwise conspicuous) object that they encounter visually after hatching. The laboratory-reared bird responds to the object much as it would respond to its parent in nature—it follows or otherwise stays close to the object and prefers contact with the first such object over other dissimilar objects to which it is later exposed.

Naive young birds deprived of contact with their own parent will behave in a filial manner toward objects that do not bear the slightest resemblance to their own species. This behavior prompted the hypothesis that species recognition in precocial birds must be a function of early learning. Specifically, it has been suggested by Konrad Z. Lorenz that precocial birds come to recognize their own species as a generalization from experience with their parents shortly after hatching. Thus, the theoretical crux of the perceptual side of imprinting is that species recognition derives from early experience with the actual parent or some surrogate that the newly hatched animal accepts as its parent on the basis of early contact. It is known, for example, that birds deprived of contact with their own species and reared, say, by humans develop a fondness for humans that in some instances precludes later social intercourse with their own species.

Most of the research on imprinting and the following-response has been concerned with the behavior of ducklings and chicks hatched in incubators in

the laboratory. The present report, however, is a distilled account of field observations made at our research station near Dorothea Dix Hospital in Raleigh,
North Carolina, over the past four years. It deals with the rapid development of
the social bond between ducklings and their parents in nature. The report concerns
mainly the behavior of two species of ducklings preceding and during their exodus
from the nesting cavity.

Among the ducks, there are two nesting types—the hole nesters and the
ground nesters. The two species we have studied most intensively are the hole-
nesting Wood Duck *(Aix sponsa)* and the ground-nesting Mallard *(Anas platy-
rhynchos).* Wood Ducks hatch out at the base of deep, vertical cavities in dead
and decaying trees, where low illumination prohibits the significant operation of
vision during the period before the departure from the nest. In the Mallard, on
the other hand, vision might be more important, as the Mallard ducklings can
see their mother as soon as she walks off the ground nest. Because of these
differences in nesting conditions in the two species, it was anticipated that the
hole-nesting ducklings might possibly be more dependent upon auditory stimulation than upon visual stimulation from the parent, while the reverse situation
might hold for the ground-nesting birds. As it turned out, however, the research
findings indicated that auditory stimulation from the parent is of prime importance
in *both* nesting types.

For our field studies, we erected some nest boxes high in trees and on posts
above both land and water, and placed others directly on the ground. The entrances
to the Mallard nesting boxes were made two times the size of the entrances to
the Wood Duck nesting boxes. (The oval entrance to the Wood Duck nesting
boxes is 4 inches wide and 3½ inches high.) In addition, the hole of the Mallard
nest box is situated close to the bottom of the box, while the hole of the Wood
Duck nesting box is located near the top, allowing a deep nesting cavity.

As it turned out, Wood Ducks always used the aerial nest boxes, while
Mallard hens always nested in the boxes on the ground. We never saw a Wood
Duck enter a ground nest nor did we observe a Mallard hen inspecting an aerial
nest.

Under typical circumstances, our method of observing the development of
the familial bond in these ducks has been to place our recording equipment and
cameras in an observation blind as far from the nest as feasible. A highly sensitive
microphone is concealed near the nest, and a transistorized recording machine
is in the blind, so we can monitor the activities within the nest box without
disturbing the hen and her brood. The magnetic recording tapes thus obtained
form a permanent record and are available for subsequent laboratory analysis by
oscillographic, spectrographic, and other techniques of audio analysis.

We begin our observations several days before the eggs are due to hatch. (The incubation period is 30 to 33 days in the Wood Duck and 25 to 28 days in the Mallard.) Our vigil begins shortly before sunrise and ends shortly after sunset, as the hen never initiates the exodus during darkness. We are interested in the kind and amount of vocal activity between the ducklings and their parent as the eggs begin to pip and the young emerge.

The exodus takes place between one and two days after the young hatch, during which time the young—sometimes as many as twenty—stay beneath the hen, receiving food from yolk sacs. The mother does not leave the nest box during this period, unless she is flushed. (If she *does* leave for some reason, she will not utter the exodus call.) It is important to emphasize that in making field observations such as these, it is essential that the observer take great care lest he alert the birds to his presence and thereby distort the "typical" procession of events by heightening the natural wariness of the birds or by frightening them.

The new (and unanticipated) information that our field studies have provided is that in both hole- and ground-nesting species the hen begins uttering its exodus call long before it leads the brood from the nest. In this way the young have an opportunity to learn the individual characteristics of their parent's call before they leave the nest (that is, before they can see their parent, as it is dark inside the Wood Duck's nest and both species are covered by the body of the mother). In figure 21.1, which presents information from a representative duck of each species, it can be seen that (1) the maternal Wood Duck begins to vocalize somewhat earlier than does the Mallard; (2) the Wood Duck utters her call more frequently than does the maternal Mallard; and (3) in both species the rate of maternal

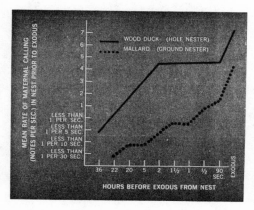

Figure 21.1. A comparison of the rates for Wood Ducks and Mallards. Note that the rate of parental calls increases as the time for the exodus nears.

calling increases throughout the entire pre-exodus period, reaching a peak during the exodus itself.

No qualitative change is discernible in either maternal call at any stage. To the human observer the basic call of the maternal Wood Duck sounds like "kuk," while the call of the maternal Mallard sounds more like "hut." A waveband analysis of the two calls indicates their audiometric similarity (figure 21.2). Within each species, there are small differences in the rate, pitch, and/or rhythm of the individual maternal calls. Between the two species, however, these differences are much greater. The difference in the rate with which the calls are uttered by the respective ducks contributes a great deal to the *perceptual* dissimilarity of the two calls (at least for the human observer).

The gradual buildup in the rate of maternal calling during the various stages leading to the exodus points to the reciprocal stimulative interplay between the hen and her brood during the pre-exodus period. That is, as the ducklings become more active inside the nest box, the hen's vocal rate increases. This is succeeded in turn by an increase in the vocal and motor excitement of the young, and so on, reciprocally, until eventually (it is my guess) the ducklings' overall level of activity stimulates the hen to leave the nest box and instigate the exodus.

Another feature of the gradual increase in the rate of the maternal call during the pre-exodus period is that such a change in rate of auditory stimulation obviates the possibility that the young might habituate to the call of the parent and not respond to it during the exodus. In the laboratory, for example, we have exposed young domestic Mallards to a recording of the high-rate exodus call before testing

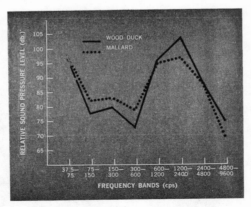

Figure 21.2. Waveband analysis of the two exodus calls shows their respective pitch frequencies and their relative loudness.

and have found it has the effect of depressing their responsiveness to it during the test, as compared to the performance of a control group that was not exposed to the parental call at all before testing. That is, the control group followed the calling model more persistently than the experimental group. "Testing" simply involves the placement of an incubator-hatched duckling inside an apparatus where the bird has the opportunity to follow a model emitting the parental call or some other call. Whether or not the animal follows the model emitting the parental call and the persistence with which it follows the model are the usual objective measures of the bird's responsiveness to auditory stimulation under experimental conditions.

The exodus in the ground-nesting species, the Mallard, is made to the accompaniment of visual as well as vocal stimulation. The ducklings, on the same level as the hen after she has left the nest, can see their parent as she calls to them. The young Wood Ducks, on the other hand, inside a deep nest high above the ground or the water, cannot see their mother after she has left the nest and must respond initially only to the parental call. (In the laboratory, we have found that both Mallards and Wood Ducks will respond to a recording of the parental exodus call. The Mallards, however, are more likely to climb out of a nest box when they can connect the call with some visual object, such as a moving striped box.) Both hens make a reconnaissance of the area before instigating the exodus, and will delay the exodus if they sense intruders.

Most laboratory workers and theoreticians have proceeded on the assumption that parental and species recognition are largely, if not solely, a function of visual cues. In nature, however, it would seem that such recognitions are founded on auditory perception. Although our field studies do not definitely answer any questions about imprinting, they do suggest that the role of auditory stimulation has been underemphasized in laboratory practice and theory. Indeed, our subsequent laboratory research, using the parental exodus calls of various nidifugous species, suggests that certain key generalizations from imprinting research may apply *only* to the visual modality and may not be germane to parental and species recognition as such occur in nature.

As imprinting is a uniquely ethological idea, and ethologists are famed for their naturalistic emphasis, questioning the correctness of the imprinting formulation on the development of species recognition in birds living in nature may seem odd. In this regard, however, it must be recalled that the "original" formulation of imprinting (whether by D. Spalding, W. James, O. Heinroth, or K. Z. Lorenz) was not based on field observation, but on the behavior of young birds isolated from contact with their own parents and reared by surrogates from other species. Thus, although the relevance of imprinting to the formation of extra-specific filial and social ties in birds deprived of contact with adult members of their own species cannot be disputed, it does seem possible that the same mechanism may not be responsible for the formation of filial and social ties in birds reared by adults of their own species.

In summary, the results of our field work suggest that in nature auditory stimulation from the parent is a typical component of the stimulative complex which, at the very least, initiates parental recognition in ground-nesting as well as hole-nesting species of ducklings. Visual recognition factors would seem to come into prominence only after the initial establishment of the parent–young bond on the basis of auditory stimulation and interaction.

22

Mother Baboon

JOAN LUFT
and
JEANNE ALTMANN

The savanna bakes in the harsh light of the long dry season and the air is still with the heat of afternoon. In the distance, groves of acacia trees seem to waver in the dusty air. A high-pitched squeaking, like an inexpert bowing of violin strings, comes form the stand of tall grass where a group of baboons are feeding, laboriously pulling the tender inner stems of the grass out of the tough outer sheaths.

The period from July to October is a critical time of the year for many animals of the East African grasslands. Most water holes are dry, the grasses are dead, and few shrubs bear fruit. At this time the baboons have to work hard for their living. They travel from morning to night in search of scattered patches of food, and they spend a long time processing the foods they do find. By the end of the day their fur is full of powdery savanna dust, and the younger animals' hands are sore from digging grass corms.

One by one the baboons emerge from the clump of grass where they have been feeding. Except for the youngest members of the group, the baboons are almost the same yellow brown as the landscape—the immense East African monochrome of dry grass and bare ground of sun and dust. The half-dozen infants born earlier in the year have black coats, wrinkled pink faces and hands, and oversized, bright pink ears that stick straight out.

As the group moves, ten-month-old Putz jumps on and off his mother's back. He blocks her way as she walks, and when she stops at another grass clump, he clings to her and interferes with her feeding. She moves him gently aside. A few minutes later, when Putz again approaches her, she stops feeding and lets him suckle. Giles—born the previous August on the same day as Putz—also tries to ride on his mother, Gin. Far less gentle than Putz's mother, Gin shoves Giles away at once and nips at him. He throws himself down, screaming and thrashing on the ground like any human two-year-old in a tantrum. From time to time he glances at Gin to see how she will react, but she is indifferent and goes on with her foraging. Finally, Giles turns to a nearby male—a formidable-looking animal, twice the size of the adult female, with a heavy mane and fanglike canine teeth. The male carries the exhausted infant to the baboons' resting place in the shade of the acacia trees.

The 45 baboons that have been scattered across the savanna all afternoon now gather in the damp green hollow under the trees. They drink at a shallow pool, mothers suckle their infants, and the older animals sit together in groups of two or three, eyes half-closed, grooming each other's fur or nibbling at leaves in the head-high tangle of bushes near the water.

Scar, an older female that has been lagging behind the other group, limps slowly toward the grove. For the last few hours she has been walking three-legged, supporting her day-old infant with one hand. She flops down in the shade with a glazed look in her eyes. Gradually, other baboons—mostly adult females and a few juveniles—gather around her, first staring at the new infant, then sniffing and pulling at it. Scar turns away uneasily, keeping tight hold of the little black infant even as another female pulls at its leg. When a male baboon comes over, the curious group drifts away, and Scar relaxes again, letting the male groom her fur as she dozes.

Dry though the savanna is—and in some places it is only bare tawny powder with animal trails scrawled white across it—water for its vast animal population is never far away. The foothills of Kilimanjaro rise behind the acacia trees, sharp and blue in the dry air, and "Kili" itself, cloud wrapped and immense, towers above them. The runoff from its snows flows under the savanna, so near the surface that the little hollows in the ground become pools and large hollows become lakes or swamps bright with water birds. The pools are also breeding grounds for malaria-carrying mosquitoes, which give the tall acacias near the water their common name of "fever trees."

Many of the trees in the grove have died—drowned at the roots in wet years, poisoned by salts in the soil, or pulled to pieces by hungry elephants. Gray, fallen tree skeletons, stumps, and scattered limbs lie in the shadow of the grove. As the infant baboons recover their energy, they make the hollow a playground. Dead trees are to climb and jump on; stumps are the place to play king of the mountain; the springy, tangled thickets of *Salvadora persica* are a maze to duck through or a trampoline to bounce on. Giles runs by with a bright feather in his teeth, like a flamenco dancer with a rose, and Putz comes running after him, eager to take the feather away.

The evening settles in (near 6:00 P.M. year-round, this close to the equator), and one by one the mothers collect their infants. The baboons climb into the crowns of three close-set fever trees, where they will prop themselves in forked branches to sleep through the night—with luck, undisturbed by the leopards that hunt in these groves, gliding from branch to branch. The two observers who have been following the baboons all day, walking with them from grass clump to *Salvadora* thicket, pack up their notebooks and their dust-covered recording gear and head back to the base camp and their own, somewhat less precarious beds.

At the whitewashed mud house that is the home of the Amboseli National Park Baboon Project, the observers go on with their work, tinkering with equipment and checking through recently acquired data. Their own supper has been much quicker than the baboons'; the evening wash water has already gone out the back door and given the grass there a few more days' lease on life. In the dark, elephants intent on their supper lumber through the yard, and as one chews loudly on the fresh grass under the window, an observer writes to the Baboon Project's American base. "Hurray, hurray, Scar finally had her kid! . . . It looks good except for some trouble clinging. Scar looks awful. Perhaps she's getting too old for this mother business. . . . The rain gauge was knocked over three times by elephants. I am holding my breath now, as it is still standing after being replaced the last time."

The rainfall records, only occasionally interrupted by elephants, are part of a much larger body of data on Scar and her companions and their environment. In the last ten years, this group of forty-odd baboons in Kenya's Amboseli National Park has become one of the most intensively studied wild primate societies in the world. It is called Alto's Group, after the regal old baboon who stood at the top of the female dominance hierarchy (the monkey equivalent of a pecking order) for many years. Alto died in 1976, but her daughter Spot has Alto's place at the top of the hierarchy, followed by Alto's other daughters and granddaughters. Researchers at the American bases, at the University of Chicago and Cornell University, celebrating the completion of a new book or article or the initiation of a new project phase, still occasionally drink a toast "to Alto—without whom none of this would be possible."

"This" includes detailed records on the baboons' environment, group membership, and behavior, on temperature, rainfall, and plant growth; on births, deaths, and such behaviors as mating, aggression, feeding, grooming, and encounters with other baboon groups and other animal species. The records on this group were begun in 1971 by Stuart and Jeanne Altmann of the University of Chicago and by Glenn Hausfater, now of Cornell University. They and a succession of their colleagues and students have kept up systematic observations and, in addition, have conducted studies aimed at specific questions about the baboons' life: the relationship of dominance rank and reproductive success, the acquisition of rank by maturing baboons, the development of foraging skills in older infants, and the like.

One of the chief focuses of the Amboseli project in recent years has been the study of mother-infant relationships. Like other monkeys and apes—and for that matter, like the humans who watch them—baboons begin life with a long period of dependence. An infant's survival depends on its mother's survival and her willingness and ability to look after it properly. Mothering is as vital a skill as foraging or defense against predators.

In southern Kenya, the harsh environment provides an acid test for motherhood. Over a quarter of the baboon infants in the study group die in their first year, and almost another quarter in their second year. (The death rate after that, when the juvenile baboons have grown stronger and are less dependent on their mothers, is very low.) While maternal care is by no means the only factor in reducing infant mortality, it is undoubtedly a significant one: in Amboseli the effects of poor mothering are sometimes painfully and incontrovertibly clear.

One of the things that has become increasingly evident in the primate research of the last few decades is that monkey mothers, like humans, do not all treat their infants in the same way. Some are indulgent, some are snappish; some quickly give in to their infants' tantrums while others do not. Some keep their infants constantly under their eye, while others let them roam unsupervised. The question naturally arises: Which of these ways of mothering is "best"? And how do the baboons—uninfluenced by either Dr. Spock or the Book of Proverbs—arrive at their various ways of treating their offspring? What makes one baboon mother "permissive" and another "strict"?

The Amboseli project's long-term records provide a wealth of data on the baboons' genealogies, environment, and past experience to help answer these questions. When the observers watch an infant growing up they know which baboons are its mother's habitual allies, which are her foes, and which are her "sisters and her cousins and her aunts"—the network of female relatives that provides much of the stability and solidarity of baboon society. They know her day-to-day relationships with males in the group, her medical history, her temperament, and her position in the dominance hierarchy. They know how many infants she has had before, how she has treated them, and how—or if—the infants have grown up. They also have comparable data on all the other infants in the group that are growing up at the same time, in the same social and physical environment.

Much of this information is impossible to obtain in the wild unless research personnel and funding are sufficient for continuous observation over many years. Generations of baboons mature and succeed each other only slowly, and Amboseli is one of the few field sites in the world where it has been possible to gather such a wide range of detailed information on several generations of free-ranging primates. On computer tapes and in notebooks and cardboard boxes, in the base camp's thatched-roof house and in laboratories at Cornell and the University of Chicago, is a rare treasure of clues to an old and unexhausted mystery: why behavior differs from individual to individual, from family to family, and from generation to generation—and what the consequences of these differences are.

Unlike the monkeys in a number of other field research projects, the Amboseli baboons are not fed by the researchers. They are thus less "friendly" with the observers and more vulnerable to shortages in the natural food supply. In this natural situation, researchers can see how the baboons cope with the harsh and prime necessity of making their own living—which, for them as much as for any other animal, dictates much of the rest of the pattern of their lives.

Far from indulging in monkeyshines, the Amboseli baboons work a Victorian factory-hand's hours. In the dry season they spend as much as 80 percent of their day searching for food and processing it: digging roots, picking berries, shelling seeds, darting after grasshoppers. Although the baboons forage together as a loose group, feeding is an individual matter—there is almost no food sharing among adult baboons—and thus a mother with a small infant has to work just as hard as a male or an adult female without an infant. If anything, she has to work harder, because as long as she is nursing the infant (that is, for much of its first year), she has to "eat for two."

Mothering thus seems to be a particularly demanding occupation, and the hard facts of Amboseli bear out this supposition. The death rate for females with dependent infants is almost twice as high as for females without infants. A mother dragging behind the group, distracted by her infant or too tired to be wary, is likely prey for a leopard or a hyena. When a fever or a viral epidemic strikes the baboon population, it is the mothers of young infants, perhaps overtired and undernourished, who are most likely to fall victim.

Scar, then, had reason to "look awful" on the day the observer saw her with her new infant. This was her fifth infant in eight years—and older mothers are even more vulnerable than young ones to the stresses of their task. If Scar had the disadvantages of age, however, she also had the advantages of experience. When her new infant cried she hitched it up so that is could reach her nipple more easily; and when it was too tired to cling by itself she supported it with one hand.

New mothers often hold their infants awkwardly for the very first few days and seem puzzled by the infants' cries of discomfort. Most learn quickly enough how to help the infant cling and nurse, but a few mothers have serious difficulties with these skills. When Vee gave birth to her first infant, Vicki, in 1976, she held it upside down and backward for most of the day, sometimes dragging and bumping it along the ground. Vicki was not able to get on the nipple until the next day. Although Vee usually held the infant right side up thereafter, she was never a very attentive or competent mother. Vicki died three weeks later, probably as a result of early mistreatment. Nazu, who had her first infant in 1979, was also clumsy and inattentive with the infant. Sometimes she sat on it. When it was six days old it was kidnapped by Scar's grown daughter Summer, but Nazu did not seem greatly disturbed. She made little effort to get it back, and three days later the infant died.

The possibility of such kidnapping is another factor, along with simple environmental pressures, that adds stress to a mother's life. Other baboons, particularly other adult females, seem fascinated by young infants: a mother is approached six to eight times as often in the weeks after her infant's birth as she was in the weeks before. This may at first seem benevolent interest, like that of humans who pause by a baby's stroller and smile. But baboon mothers are often nervous of their neighbors, and they often have reason to be. Another adult female's "interest" in the infant may extend to snatching it and carrying it away with her. Since she cannot nurse it and may mistreat it, the result, as in the case of Nazu's first infant, may be death.

It is not only the offspring of indifferent mothers that are subject to kidnapping. Handle was one of the two most protective mothers in the entire group, but she lost her two-day-old infant, Hans, to another adult female and, despite her persistence, was not able to get him back until the next day. By then he was dehydrated and weak, unable even to cling to his mother. Unlike Nazu's infant, however, Hans recovered from the kidnapping. Hans's kidnapping points up an important fact about maternal care; namely that the mother and her behavior are not the only factors in the infant's life. A very protective mother may be overpowered by stronger individuals; an indifferent mother's infant may get care from the rest of the group.

After its mother, one of the most important figures in the infant's life is an adult male that is a frequent companion of the mother and that is often, but not always, the infant's father. This special male companion lets the infant ride on him, sit in his shade on hot afternoons, and eat scraps of his food when it is old enough. Equally important, he protects the infant from too-curious neighbors. Other baboons are less ready to approach a mother and a new infant if their male protector is with them, and he may overtly threaten animals that try to pull at the infant.

Developing a helpful relationship with an adult male, like caring for the infant itself, is not something first-time mothers do immediately or without difficulty when their infant is born. Handle, like many new mothers, took more than a month before she accepted and made use of adult male Even's attempts to stay near her. It was during this initial period, when Handle was still going it alone, that Hans was kidnapped—and one of the reasons she could not get him back more quickly was that the kidnapper, Gin, enlisted the aid of her companion, Red, a young adult male just beginning his rapid rise through the dominance hierarchy of the group. He helped keep Handle from getting close enough to Gin to retrieve Hans.

Gin gave birth to her own first infant about three months later. She was far less protective of Grendel than Handle was of Hans; but, in a sense, she had less need to be. Although she was as inexperienced as Handle and about a year younger, she was somewhat higher ranking and also made immediate use of her relationships with males. Red stayed close to Gin and Grendel from the first and kept the other baboons from pulling at Grendel. (There was no evidence that he was Gendel's father.) He had also approached Handle and Hans in recent weeks, with the apparent intention of playing his protector's role there, but Handle, perhaps remembering his part in the kidnapping, avoided him. By then Hans and Handle had acquired a male companion of their own that helped keep Red away.

When Grendel was a month old, Red left the group. It is common for young males to move in and out of groups, and Red was a particularly frequent migrant during this year. His weeks with the group were full of fights with other males, as he worked his way up the dominance hierarchy. His first absence after Grendel's birth was one of his longest—nearly two months—and he and Grendel never resumed their early relationship after his return. Grendel had meanwhile picked up a new protector: his probable father, High Tail. The two of them were particularly close, in large part because Gin tended to reject and avoid her infant. Grendel sat next to High Tail, rode on him when the group moved, and ran to him in moments of alarm or distress. In May, at the end of the long spring rains, an epidemic struck the group. High Tail fell very ill: for some days he did not eat. He moved as little as he could and swayed when he moved. Two baboons died in this epidemic—typically enough, a mother and her young infant—but by the middle of the month High Tail was recovering.

One morning the observers arrived to find the whole group agitated, apparently by the presence of a leopard in the high grass around the grove. The baboons came down from the trees later than usual and moved away quickly in a tense compact group, quite unlike their usual, scattered, dawdling morning start. As they left the grove, several juvenile baboons looked up into a nearby tree. The observers, following the baboons' line of sight, saw High Tail's half-eaten body hanging on a branch—apparently, he was a victim of the leopard that waited and watched in the long grass.

Grendel, now four months old, was again without an adult protector. He tried to spend more time with his mother, but Gin was unwelcoming as before— shrugging him off when he tried to climb on her for a ride and snapping at him if he grew insistent. He tried to attach himself to a number of baboons in the group, but none was as tolerant of him as High Tail had been.

Baboons seem to make a distinction between very young infants with black coats and older infants with adult coloring. It is the black infants that they gather around and want to touch, and it is the black infants that they are usually ready to carry. Grendel's coat was turning golden brown when High Tail died, and although he got brief rides from many members of the group, he had to survive largely on his own. He still nursed from Gin, but only in her more tolerant moments.

Hans, on the other hand—two months older than Grendel—could suckle and ride on his mother almost at will. Handle, anxious and protective, still let him stay close to her. Even Handle, though, eventually hit Hans and shook him off when he tried to ride, and even Gin had let Grendel suckle at will when he was very small. The difference between "harsh" and "gentle" mothers is perhaps not so much an absolute difference of temperament as a difference of timing. Grendel's mother first rejected him when he was one month old; Hans's mother not until he was five months old.

The difference between care and rejection is a matter of timing in another sense, too. As the infant grows larger, the mother needs to feed more to produce enough milk. At the same time, the infant is capable of greater independence and can play by itself without risk of kidnapping or of getting lost. Thus when it is very small, the infant rides on the mother, suckles from her while she is feeding, and moves away to play only when she has stopped to rest and can keep an eye on it. By the time the infant is four or five months old, it is seriously in its mother's way if it rides or suckles while she feeds, and she cannot afford to let it interfere. If she is malnourished, the infant will be too. At this age, the best arrangement for both mother and infant is for the infant to run about on its own while the mother is busy foraging, and for the infant to cling to her or suckle only when the mother sits down to rest—an exact reversal of the daily pattern of earlier months.

The accomplishment of this reversal seems to cause the most trouble between baboon mothers and infants. The fourth month of the infant's life, when the transition usually takes place, is rather like the "terrible twos" for human children. Infants throw violent tantrums, and a mother's patience wears thin. (In the first few months of their infants' lives, Spot and Gin were the only mothers that acted aggressively or punitively toward their infants. By the end of the fifth month, however, every mother in the study group had bitten, pushed, grabbed, or hit her infant.)

The question is why some mothers push their infants toward independence at an early age while other mothers hold them back until later—and which method is best. The key to the question, as it turns out, is maternal rank. Whereas male baboons change rank fairly often, rising and falling in the dominance hierarchy, most females keep the same rank throughout their adult life. They also stay in the same group, whereas the males migrate. Thus the stable core of a baboon society is its adult females, which range together over the same area for twelve or fifteen years and maintain the same relationships with each other, while males come and go and infants are born and die or grown to adulthood around them. As juvenile females grow up, they usually take a place in the hierarchy just below their mother's, so that the stability of female baboon ranks continues across generations.

In baboon, as in human, society, life is rather different for individuals at the upper and lower ends of the hierarchy—and this is as true of motherhood as of any other aspect of life. All but one of the higher-ranking mothers in the study group were early rejecters of their infants, while all but one of the lower-ranking mothers were protective. Alto and her daughter Spot, for example, at the top of the hierarchy, pushed their infants toward independence at an early age, while Brush and Handle, near the bottom of the hierarchy, kept their infants "tied to their apron strings" for months.

And which strategy proved best? In the Amboseli study it was evident that one strategy was best for some baboons, while the other was best for other baboons. Even within one small social group, the circumstances in which mothers found themselves varied greatly; and the way they treated their infants varied to fit the circumstances. For infants of higher-ranking mothers, the social world is relatively benevolent. Other baboons are potential helpers—or can be made to be helpers. Alto's daughter Alice spent much of her time with other mothers in the group, sitting in their laps and even pushing their infants aside in order to make room for herself—thus earning the sobriquet Alice the Obnoxious from one observer. If the other mothers pushed Alice away too roughly, they were likely to get trouble from their "superiors," Alice's mother and sisters.

Alice's relative independence from her mother proved to be beneficial not only to Alto but to Alice herself. When Alto died, Alice was only sixteen months old—the youngest orphan in the Amboseli study ever to survive alone. It is possible that if she had been less accustomed to feeding and traveling on her own, and less practiced in enlisting the aid of others in the group, she would not have survived her mother. For the lower-ranking mothers, on the other hand, the social environment is hostile. Other baboons are potential kidnappers, food stealers, and attackers of infants. A mother at the bottom of the hierarchy needs to be protective. For her infant, the risks of early independence are likely to outweigh the advantages.

Since young female baboons like Alice "inherit" their mother's rank, while young males like Grendel and Hans have to fight their way to whatever rank they can reach by their own exertions, it is peculiarly suitable that high-ranking mothers in the Amboseli study have tended to bear more female infants, which are sure to be high ranking themselves, while low-ranking mothers have tended to bear more male infants, which at least have a chance to work their way to a higher rank. By some process that is not yet understood, mothers seem likely to give birth to the sex of infant that will have the best chance in life.

The sex-ratio difference is a subject for continuing research at Amboseli, as is the behavior of aging mothers such as Alto. The observers did not know Alto in her youth and had no opportunity to compare her earlier and later maternal behavior. But by now some of the females that were hardly more than infants themselves when the Amboseli project began have had several infants of their own. It remains to be seen whether, as expected, they will be less restrictive with their new infants than they were with their earlier offspring. As a mother grows older, the likelihood increases that she will die before her infant, as Alto did. It would be to the advantage of the infant to learn to survive on its own at an earlier age, and for that reason (if not from sheer exhaustion) the mother may watch over it less closely than she watched over its older siblings.

The infants that were born in the first year of the mother-infant study have just become mothers, and it remains to be seen whether, as expected, they will bring up their infants in the same way they were brought up. The mother-infant study turns to new sets of questions; meanwhile, systematic observation of Alto's Group goes on from season to season.

In November the first rains come and the savanna turns green. Soon the fever trees and their relatives the umbrella trees come into flower, and the groves are hung with puffball clusters of white and yellow blossoms. The infants born in early summer are beginning to find their own food and eat the sweet flowers like candy. The umbrella trees especially, with their rough bark and low branches, tempt the infants to climb. When the group comes to one of the umbrella-tree groves, whose flat, feathery lines mark the savanna horizon, the infants go scrambling up, eat their fill of sweets, and swing Tarzanlike down the vines that hang from branch to branch. The open land is carpeted with the pink-lavender, primroselike flowers of *Rhamphicarpa montana,* which the infants pull by the handful to eat. Sometimes a small baboon, its mouth still full of flowers, comes bobbing across the grassland, eager to join the group. At this time of year life is easier for the baboons, and they need less of the day to find and process food. Most of the next year's infants are conceived at this time, the relaxed and full-fed days of the warm-season rains.

It is the best of times and the worst of times—for baboons and scientists, respectively. "This is not my conception of doing research," one new observer, undergoing his baptism of mud, wrote home. "Dec 17 it rained and we could not find any baboons; Dec 18 it rained but we found Alto's Group, however we used up half a tank of gas driving in 4-wheel and had to leave at noon to avoid being stranded; Dec 19 it rained . . . spent most of the day digging ourselves out of the Emali road; Dec 20 it rained and we could not find any baboons; . . . [addenda Dec 20; 1 flat tire during the day, that night organized search party to find (other researchers) who were 3 feet deep in mud]; Dec 26 finally found Alto's Group, the steering control on the Land-Rover broke, we abandoned the vehicle in the field. . . ."

With a little practice, however, the baboons are easier to find, and the harder soils on the savanna easier to tell from the bottomless mud. In January and February comes the respite of the interrains. In March the long stormy season sets in; Kilimanjaro's mantle of snow hangs lower, and cloudbursts sweep the savanna. By the time the worst mudholes have dried out again, the first of the infants conceived in the "shorts rains" will be born.

"Bring on the cigars!" runs the message from Kenya to Chicago. "Vee had her baby! Another girl, as one would predict." Vee is high ranking, and so far has produced nothing but female infants. "Voodoo has crossed eyes but otherwise is looking pretty good." Vee is more experienced now than she was five years ago, when she bumped the unfortunate Vicki like a potato sack across the savanna: "Vee has handled her very well; even Alice the Obnoxious has to watch her p's and q's." The information about Voodoo's birth is filed in a notebook and entered in computers at Cornell and Chicago, where it can be matched up with dozens of other sets of data on births and deaths and matings, aggression, feeding, neighbor relationships, and group movements. One of the Chicago workers—following the Amboseli news like the latest installment of a soap opera—calls down the hallway, "Hey, Vee had her kid!"

If Voodoo is left more on her own than her older sisters were or if she is treated differently from infants born at different times of the year or if her mother expends more energy on her than a restrictive mother like Handle, the researchers in Chicago will know, in this new phase of the mother-infant study—and we will all know more about the effects of mothers and infants on each other's lives and about their relations with the society around them.

23

A Mound of One's Own

W. THOMAS JONES and BETSY BUSH

Travelers of back roads in southern New Mexico, southeastern Arizona, or northern Chihuahua and Sonora in Mexico may notice the broad, low earthen mounds scattered over the desert flatlands. Anyone who stops to examine them will find that each mound has numerous holes that seem to be entrances to tunnels or animal burrows. However, a daytime visitor who hopes for the animal occupants to emerge will inevitably be disappointed; they only come out after dark. On moonlit nights one can see them—creatures, a little bigger than hamsters, that hop on two legs like kangaroos and have long white-tipped tails. These last two features inspired their name—banner-tailed kangaroo rats.

Bannertails are one of fourteen kangaroo rat species in western North America. Most of these species are found in the arid regions of California, Nevada, and Arizona, although one species' range extends from Canada to Mexico. Kangaroo rats are members of a rodent family, Heteromyidae, that includes two other desert genera—pocket mice and kangaroo mice—and several tropical species. The kangaroo rats are believed to have evolved into a distinct group as land upheavals and mountain building caused the formation of deserts in western North America about seven million years ago. Within their family, kangaroo rats are among the most highly specialized for life in desert environments. They are nocturnal, live in underground burrows, are bipedal, and have long tails for balance. Their small forelimbs, which look more like hands than feet, are used mainly for digging and for stuffing food into pouches in their cheeks. Food in this case means seeds, supplemented with green leafy plants. Some kangaroo rat species, including bannertails, have an interesting means of communication. They drum their hind feet rapidly on the ground, often when other kangaroo rats intrude upon their territories or burrows, and sometimes in response to intruding biologists.

One of the greatest hardships of desert life is the lack of water, but kangaroo rats can survive without ever drinking. They get all the water they need from the seeds and green vegetation they eat. How they manage is a study in economy. First, they lose very little water in excretion; their efficient kidneys produce a highly concentrated urine. Second, their convoluted nasal passages condense most of the moisture out of their exhaled breath before it reaches the dry desert air. Finally, what water they do lose through their breath they recover in their burrows, and during the day they seal off the entrances with soil. The moisture in their breath and in the soil humidifies the burrow interior, and some of that moisture is absorbed by the stored seed. Later, when the rats eat the seed, they recover some of their lost water.

As interesting as these physiological adaptations may be, it is the social system of bannertailed kangaroo rats, and what it tells us about the evolution of mammalian social behavior in general, that has been of particular interest to us. Bannertails are considered a solitary rather than a gregarious, species, but certain aspects of their natural history, particularly their large mounds, have made us suspect they might have some gregarious tendencies. By studying the rudiments of gregariousness in a basically solitary species, we hoped to learn more about why a species might start down the evolutionary path toward gregarious sociality.

Since the summer of 1979, we have been studying bannertails in a remote valley in extreme southeastern Arizona. The valley supports a dense population of rats, along with sparse shrubs, such as mesquite and Mormon tea, and various annual plants. These are the principal sources of seed for the kangaroo rats. Our 90-acre study site has more than 300 individual bannertail mounds. After trapping and marking each kangaroo rat in our site, we have continued (most recently in collaboration with Peter Waser of Purdue University) to census the study population every two to four months by trapping at each occupied mound. By noting which individuals are captured at each mound, we have been able to learn each rat's home mound and have identified the mothers of most juveniles. We have also monitored some rats in more detail using radiotelemetry. By compiling lifetime records of the movements and locations of every rat, we have discerned much about their social system.

What distinguishes gregarious species from solitary ones? Many highly gregarious mammals live in extended family groups. Black-tailed prairie dogs, for example, live in groups called coteries that include several females, their young, and one or two males. Members of a coterie share a territory and its extensive set of burrows. Because young females almost never disperse and eventually replace their mothers as the breeding residents of their natal coteries, females in a coterie are closely related—they are sisters, half-sisters, aunts and nieces, or mothers and daughters—and coterie territories tend to stay in the same lineage. Within coteries, prairie dogs get along amicably, but coterie territories are aggressively guarded from noncoterie members.

In solitary species, on the other hand, males and females usually live alone and defend their territories against all other individuals, including relatives. In addition, the young usually do not spend their entire lives in their natal territories unless they inherit them when their parents die. The Merriam's kangaroo rat is a good example of a solitary mammal. These rodents always live in their own burrows and the young disperse soon after they are weaned, so family groups persist for only a short time. As adults, kangaroo rats respond aggressively to all individuals that intrude on their territories.

These two examples lie at opposite ends of a spectrum of sociality that includes many intermediate states. Some gregarious animals have more loosely organized social systems than the black-tailed prairie dog's, while some supposedly solitary species show gregarious tendencies. There are also gregarious animals that live, not in extended families, but in groups of unrelated individuals. Nonetheless, extended families are certainly the most common social grouping in mammals; they are the basic social units in species as diverse as elephants, naked mole rats, and baboons.

Most researchers believe that gregarious social systems evolve from solitary ones. One way this might occur is if a species begins to extend the period that offspring live in natal burrows or territories, thus extending the time they are in close association with relatives. If this tendency evolved to the point where offspring never dispersed, close-knit family groups would become permanent, and the system would be considered gregarious.

But what could cause a delay in the dispersal of offspring? The young might prolong their stay in natal areas if that would give them a reproductive or survival advantage over dispersing. If, for example, there are essential resources in the natal territories that would be unavailable to juveniles that left home, then dispersing might be risky. Female thick-tailed bush babies, African primates, probably remain at home for just that reason. The young females need access to gum-producing acacia trees if they are to have enough food to be able to reproduce. When Ann Clark of the Kellogg Biological Station in Michigan studied bush babies, she found that adults monopolize all the trees and only allow their own young to feed on them. Thus, the only way females can get at the trees is to stay in their mothers' territories.

Sometimes a scarcity of suitable habitat may be the critical factor. Jerram Brown's study of Mexican jays in Arizona suggests that since most of the breeding habitat on his study site is taken by adults, young jays stay in natal territories for several years. Eventually, nearby territories become vacant or juveniles may ascend to breeding status at home. Parental tolerance of late dispersal by the young may be a form of prenatal care that increases the young's chances of survival.

With these ideas in mind, we began to study the mounds and dispersal behavior of bannertailed kangaroo rats. Bannertail mounds are so large and complex it seems remarkable they are inhabited by an animal as small as a kangaroo rat. Each adult rat lives in its own mound. Some mounds are as much as fifteen feet in diameter and two feet high. The tunnel system inside runs as deep as two feet below ground. Usually these mounds have a dozen or more entrances, which make them look like rat condominiums. Inside are passageways that wind, split, and converge again in fragile, underground labyrinths.

A burrow system such as this requires a long time to build. A rodent takes months or even years to expand a simple, single-entrance burrow into a small or average-sized mound. Perhaps this is why bannertails rarely build new mounds— about three new ones are constructed per hundred each year. Most of the mounds on our study site are probably decades old, much older than the rats themselves, which live at most three to four years.

Mounds are crucial to bannertail survival. They provide refuge from extreme desert temperatures and aridity, as well as from predators such as snakes, owls, and coyotes. The mounds also serve as granaries. Most seed-producing plants of the area bloom and set seed after summer rains; bannertails harvest these seeds in autumn, storing up to eleven pounds per mound. Their caches sustain them during the winter and spring months when food is relatively scarce. Once accumulated, the caches require constant maintenance because the seeds tend to become moldy. Jim Reichman and Cindy Rebar of Kansas State University have found that bannertails prefer slightly moldy seeds to nonmoldy ones, probably for nutritional reasons, but they don't eat very moldy seeds. When Reichman studied this in the laboratory, he found that kangaroo rats appear to manage mold growth by moving seeds to areas of different humidities.

Each juvenile bannertail must acquire a mound if it is to survive to adulthood. But the time and effort required to construct new mounds probably prevent most young from doing so; juveniles must usually find existing mounds. Most mounds, however, are occupied, so juveniles can only live in their mothers' mounds. Furthermore, young bannertails may need access to seed caches during their first few months of growth. Given these circumstances, we predicted that young bannertailed kangaroo rats should delay dispersing from their mothers' mounds and leave only when vacancies appear elsewhere.

We found this is exactly what they do. Each year in January or February most females one year or older enter estrus and produce one or two litters by late spring. The mothers nurse for about a month; but after weaning, offspring do not immediately leave home. Instead, virtually all young of both sexes stay with their mothers for at least another two months. Most remain for four months, and some remain for as long as seven months. This is substantially longer than the usual pattern for solitary mammals, whose juveniles usually leave soon after weaning. In one case, a young bannertail and its mother shared a mound for eighteen months, at which point the mother died.

We discovered that when young bannertails finally do leave their natal territories, they usually move to adjacent mounds, and these moves always follow the disappearance of the previous resident. We have never seen a juvenile displace an adult. In several instances when we knew an adult rat had died, we found that a juvenile moved in within days. This pattern suggests that dispersal is prompted by the opportunities that arise when adult kangaroo rats die.

Not all juveniles leave their maternal territories and take up residence in nearby mounds. Some acquire mounds in natal territories. Many adult females have extra mounds in their territories and sometimes juveniles resettle in one of these. It is also not uncommon for an adult female to relocate herself, thereby relinquishing the natal mound, and whatever seed it contains, to her offspring. Offspring may also inherit natal mounds when their mothers die. Our data show that those that acquire natal mounds are more likely to survive to adulthood than those that move elsewhere. Juveniles that acquire natal mounds usually remain there all their lives and eventually pass these mounds on to their own progeny. Hence, a mound or group of mounds may remain within a lineage for several generations.

This pattern of delayed dispersal leads to the formation of rudimentary social groups among bannertails. In spring, females sometimes give birth to second litters while still sharing mounds with their first litters. At this time, as many as a half dozen closely related individuals may share a single mound, and occasionally daughters of early litters begin to reproduce. Such groups seem to resemble the extended families of prairie dog coteries, but we know little about their social dynamics. And unlike prairie dog coteries, bannertail family groups do not persist indefinitely. Mortality is high among juveniles. Many die before or during dispersal, and the majority do not reach reproductive age, which is usually one year old. Moreover, mothers may not tolerate offspring at home after the beginning of the fall seed-caching season because they must accumulate and save a new store of seeds for the following year and the next litter. Finally, the shortage of suitable mounds does not last forever. Juveniles take advantage of openings, and by the age of seven or eight months, most are independently established. Although the existence of family groups is only a transitory phenomenon among bannertailed kangaroo rats, our study suggests that incipient social groups such as theirs could become more permanent if conditions change so that critical resources in natal territories are at a greater premium. For example, lower adult mortality rates among bannertails would decrease the rate at which mound vacancies arise.

In all plants and animals, parents provide their young with resources to help them become mature and independent. The endosperm of seeds, the yolk of eggs, and milk are all things that offspring in their undeveloped state cannot obtain themselves. When juvenile bannertailed kangaroo rats stay in the mounds of their birth several months past wearning, they may be receiving a similar benefit. Young bannertails are unable to compete with more experienced adults for existing resources, so parents provide them with access to a mound and seed cache. Thus, permanent nondispersal and its consequence, gregarious sociality, may be a highly developed form of a very general phenomenon, parental investment in offspring.

24

Sea Lion Shenanigans

FRED BRUEMMER

Various hazards threaten the lives of Hooker's sea lion pups during their first weeks of life. Some are crushed by huge sea lion bulls rampaging across the beach in pursuit of rivals, a few drown in steep-sided pools, and others lose contact with their mothers and starve. On the remote sub-Antarctic island of Enderby, northernmost of the Auckland Islands, 300 miles south of New Zealand, the pups face a unique danger—10 percent of all sea lion pups born on this island die in the burrows of feral rabbits.

More than a century ago, sealers, settlers, and explorers released rabbits of a French breed, Argenté de Champagne, on Enderby Island. Now about 4000 rabbits live on the island, and part of the dune and sward area above the sea lion rookery is honeycombed with their burrows.

Sea lion pups are curious. They also like the feeling of being touched and surrounded. Rabbit burrows intrigue them. A pup wriggles into a burrow. Another follows, nipping at the flippers of the first one. They squirm farther and farther into the dark, warm encompassing tunnel. It is a one-way trip; the pups can advance for a while, but they cannot retreat. When they become stuck, they struggle—fiercely at first, then feebly—and finally suffocate.

Martin Cawthorn of New Zealand's Fisheries Research Division, head of the 1980–81 Auckland Island Expedition, discovered the first dead pup on Enderby in a rabbit burrow on January 10, 1981. From that day on, until we left the island in February, we found entombed pups nearly every day. Cawthorn pulled seven dead pups from a single burrow and five out of another.

With a total population of about 7000 the Hooker's sea lion is the rarest, least known, and geographically most restricted of the world's five sea lion species. A few breed on Campbell Island, 400 miles south of New Zealand. Three other rookeries are in the Auckland Islands. At least 2000 sea lions haul out on small, reef-ringed Dundas Island (200 yards by 150 yards), which in some areas becomes so crowded that, as New Zealand naturalist Sir Robert Falla observed in 1978, the pups lie "head to tail like canned sardines and sometimes more than one layer thick." Another 100 breed in the dense forest of tiny, lush Figure-of-Eight Island. And 600 breed on Enderby, almost all on the broad, nearly half-mile long beach at Sandy Bay on the island's south side.

The Auckland Islands were discovered in 1806 by a British whaling captain, Abraham Bristow, and within a few years sealers of many nations, from the United States and Britain, from Australia, and from France were killing the islands' valuable fur seals and, as a sideline, the less valuable sea lions. Captain Benjamin Morell of the American Merchant Service wrote: "Although the Auckland Islands once abounded with numerous herds of fur and hair seals [sea lions] . . . [the sealers] have made . . . clean work of it." In a detailed survey he made of the islands in January 1830, at the height of the sea lions' breeding season, he saw only twenty sea lions and not a single fur seal.

This lack of fur seals and sea lions halted the hunt on the Aucklands, but whenever the animals' numbers increased a bit in the following decades, sealers returned and renewed the slaughter. In 1865, a New Zealand government ship searching for castaways on the Auckland Islands reported "plenty" of sea lions. When the final hunt took place in 1880, the crew of the New Zealand sealer *Friendship* was kept busy killing sea lions on Enderby Island. In 1881, the Hooker's sea lion received total protection from the New Zealand government, and in 1934 the entire Auckland Islands archipelago was made into a reserve for the preservation of fauna and flora, now administered by New Zealand's Department of Lands and Survey.

Sandy Bay on Enderby, scene of so many sea lion massacres in the past, was serenely beautiful when we arrived in November of 1980. Rata forest, hemmed by wind-sculptured thickets of myrsine and cassinia scrub, covers about half the island. Between the forest and the beach, a broad expanse of sward is kept lawn smooth by the islands's wild cattle and the 4000 busily nibbling rabbits.

The beach's central portion belonged to sea lion bulls in their prime, at least eight years old, and weighing between 700 and 900 pounds (318–408 kg). This was the beginning of the breeding season, and the bulls were fat and sleek. The belly fur of a Hooker's sea lion bull is a warm, chestnut brown, its dorsal pelage is deep gray-brown, the long mane fur, which envelops the powerfully muscled forequarters like a triangular cape, is crisped and slightly grizzled, the head and muzzle are nearly black. The bulls are massive, majestic, and when they glare at rivals with watery, bloodshot eyes, rather menacing. They haul out on the beach in early November, at the beginning of the southern summer, and after some vicious fights and much posturing, divide the beach among themselves, each bull occupying a territory of from 12 to 25 feet in diameter.

On the seaward side of the territorial males, "beach bulls," spaced at roughly 15-yard intervals, occupied the water's edge, advancing and retreating with the rise and fall of the tides. Most were subadult males, about five to eight years old, as long (about eight feet) but not as massive as the territorial bulls. The beach bulls do not hold territories, so they try to intercept females on their way to and from the sea.

On the landward side another cordon of subadult bulls, at the edge of the beach and just beyond it on the sward, had similar ambitions. They attempted to waylay any landbound female. For nearly a month they all were bulls in waiting, for no females had arrived on their beach.

While the mature bulls spend weeks defending private circles of sand on a beach occupied solely by their own sex, the females mass three miles away at the southeast point of Enderby Island. This division has obvious advantages. When the females finally come to Sandy Bay to bear their pups, most territorial disputes between bulls have been settled, neighboring males know each other, and the exact limits of their respective domains, and disruptive fights between bulls are neither as frequent nor as fierce and prolonged as they are in the early stages of territorial acquisition and defense.

Compared with the massive males, female Hooker's sea lions seem small, graceful, and svelte. Most weigh between 180 and 250 pounds (82–113 kg). Head, neck, and back are grayish fawn, flanks and chest cream colored or sand yellow, the belly fawn or sienna. They are intensely gregarious. On a beach, it is not the bulls that attract females but the presence of other females. They usually lie together as tightly packed as possible. They often bicker, but never bite each other.

At the beginning of December, about 350 females had assembled at the "staging area" at Enderby's southeast point. They formed a nearly contiguous mass, and packed in among them lay more than 50 males. Some of these bulls controlled territories and these were now covered with females. But the reign of the bulls is ephemeral. Females come and go as they please, and within three weeks all but 19 had left. These 19 seals bore their pups on the staging area and remained there during the entire breeding season. According to Cawthorn, who has spent previous seasons at Enderby, this was the largest number of pups he had seen born at the southeast point and it seems possible that a new breeding rookery is beginning to establish itself at this place.

Many of the females at the staging area were accompanied by the previous year's pups. These hefty yearlings still nursed frequently and then slept, pressed to their mothers, or played with peers at the rookery's edge. When a yearling is hungry, it cries loudly. Its mother wakes and answers, and the youngster squirms across the mass of resting females, finds its mother, and starts to nurse.

In early December, the females began to abandon the staging area, and after a five-day delay they arrived at Sandy Bay. During these days, presumably spent feeding at sea, the females shed their yearling pups. Only one hauled ashore with its mother at Sandy Bay. We frequently encountered the abandoned yearlings. Some slept in the broad meadows that gird the island. A few walked far into the forest. For a while we could hear them call mothers that were no longer there to respond. Then the year-long bond broke. The yearlings ceased their plaintive wailing, slept alone with other abandoned yearlings, and after a few weeks returned to the sea.

It is not easy for the females to come ashore at Sandy Bay. To reach the beach area, they must run the gantlet of beach bulls that line the water's edge in eager anticipation. Females, as a rule, arrive singly. A female often swims back and forth a few times near the beach and then suddenly rushes ashore. Immediately, all the beach bulls in the vicinity converge upon her. The first to reach her tries to pin her down, but she usually repels or eludes him. She may dash back into the sea or she may flee landward, galloping across the sand with amazing speed. Often the bulls themselves confound their purpose. They are so jealous and so busy fighting each other that the object of all their attentions has ample opportunity to escape.

Once the female reaches the domain of a territorial bull, she is "safe." Unless she is in estrus, the resident bull will not bother her and his section of sand is off limits to other males. The first female came ashore on December 6 and settled on a territory in the middle, and toward the eastern end, of the beach. What made her choose this spot, we do not know. But once established, she became the magnet for all subsequent arrivals. Neither bulls nor territories, it seems, hold much allure for the females. The presence of other females is what attracts them. Thus, on a broad beach, half a mile long and dotted with a hundred bulls, the arriving females at first clustered tightly upon the territory of a single male. By her choice, the first female had determined the location of the breeding rookery at Sandy Bay and the fate of its resident bulls. Some bulls had females galore, most had none.

While females value females and seem to regard the males as a sometime threat and all-time nuisance, the males treasure territory and appear to regard females, if they are present upon their territories, as an incidental bonus. Territorial bulls make no effort to acquire females and if any happen to be within their domain, they only rarely, and even then rather half-heartedly, try to retain them, so that females can move freely from one territory to another.

Only nonterritorial bulls actively pursue females. Territorial bulls are rooted to their private space for more than two months, during which time they neither eat nor drink. And they will defend their territories with equal vigor against all other adult males regardless of whether there are any females on them or not.

Four days after the arrival of the first females, pups were born at Sandy Bay and the structure of the rookery changed rapidly. At this time 47 sea lion cows had come ashore and the tight initial cluster of females began to spread across the territories of several bulls. This was caused in part by the arrival of more and more females (392 finally assembled at Sandy Bay), but mainly because just before, and for about a day after, parturition, the females evince a need for privacy and space.

Normally friendly and almost compulsively gregarious, a female that is about to give birth becomes tense and hostile. She twists, thrashes, and circles, snarls and snaps at neighbors. The territorial bull, which does not tolerate quarrels among cows upon his property, lumbers over, interposes his bulk between the feuding females, and grunts angrily. The female resumes her circling and thrashing. Slowly she creates an empty space around herself. Other females in labor produce similar private spheres, forcing nearby females to move and, in a ripple effect, jostling others, until peripheral cows spill over onto the territories of adjacent males.

The birth of a pup takes from fifteen minutes to two or three hours. The dark, glistening newborn pup glides into the churned-up sand and lies quietly for a few moments. Arching backward, the female sniffs the pup, then opens her mouth wide to lift and hold the pup for an instant. The pup cries, a tremulous, lamblike bleat. It stirs and pulls itself feebly forward with its front flippers. The female twists her abdomen abruptly, yanking the pup backward and breaking the umbilical cord. A horde of eager southern skuas surround the female and her pup. The boldest one sneaks in and peeks at the pup's trailing unbilical cord. The female snaps, the bird retreats. The pup raises itself a little on its foreflippers, shakes its big head, and topples over. Leaning sideways, the female fondles the pup with her long front flipper, rubs it, sniffs it, and listens attentively to its bleats. Henceforth, her pup's distinctive smell and sound will enable her to identify it from among all the other pups on the beach.

The pup now begins to search for its mother's teats, sniffing and snuffling along her body. She reclines. The pup mewls and mumbles and makes little sucking noises. Finally, it finds one of the four teats and begins to nurse greedily. Droplets of milk trickle from the corners of its dark muzzle. About twelve minutes have elapsed since its birth.

For two or three days, the newborn pup stays close to its mother. Most of the time it sleeps, pressed to its mother's body, waking only to nurse. During this initial phase the female is very solicitous and protective of her pup. She may cradle it with her flipper, and she sniffs it often. After about two days the pup shows signs that it is beginning to recognize its mother. If it moves a bit away from her and she calls, the pup responds, although at first this vocal recognition appears to be vague and poorly focused.

The most disruptive elements in the rookery are the three- to five-year-old immature males. About the size of adult females, these young males are active and feisty, venturesome yet fearful. They spend much of their time play-fighting with peers on the rookery's periphery. But many insinuate themselves among the females. As long as they keep a low profile, the territorial bulls, as a rule, tolerate them. But when such an enterprising youngster begins to court a female and attempts to mount her, she angrily and loudly rebuffs him. Alerted by the noise, the resident bull charges and the young swain flees. If he is cornered by an adult bull, he usually assumes the submission posture of a female. This seems to confuse the big bull, and the young interloper escapes unscathed. But the endless shenanigans of these young males, as well as incursions by subadult bulls and quarrels with neighbors, incite the territorial bulls to frequent charges across their territories, thus endangering the newborn pups.

Adult bulls are not actively hostile to pups. They simply ignore them, treating them as if they were but flotsam on the beach. If a pup is in a bull's path, he tramples it. A pup weighs about 12 pounds (5.4 kg) at birth, and its bones are soft and pliant. Even when steamrollered by a 900-pound bull, the pup usually escapes without injury. And, within two to three days after birth, a pup is fully aware of this threat and flippers frantically out of the way when one of the colossal bulls bears down on it.

The greatest danger is that a bull will abruptly halt his charge while squarely upon a pup. If the bull just sits there, the tiny pup will struggle feebly but futilely and, unless the bull moves, will eventually succumb. Most females try to shield their pups from this menace. When a male charges in their direction, they place themselves protectively between the bull and pup. Some grab their pups and yank them out of harm's way.

If a female's pup is pinned beneath a bull, she becomes extremely agitated but not once in such a situation did we see a female attack a male. If any part of her pup protrudes beneath the bull's bulk, she may grab it and try to pull it out. When the bull moves, freeing the pressed pup, the female sniffs it worriedly and occasionally will lead it to a different section of the rookery. Cawthorn watched one female use a shrewd stratagem to free her pup. She nibbled the bull's neck, as cows do who are in estrus. Then she turned provocatively and presented herself for mounting. The bull moved immediately, she rebuffed him, lunged past him, and rescued her pup. Despite such maternal wiles and precautions, approximately 3 percent of all pups born at Sandy Bay are crushed to death by bulls.

When they are about three days old, the pups start to wander and explore. They stare as gulls and skuas attempt to chase them, trip over their long flippers, and land headfirst in the sand. They also begin to play with other pups. Most are nimble enough by now to escape from the galloping bulls, and they have nothing to fear from the females, which are tolerant and gentle. Females may threaten a strange pup, but they never bite or maul it.

Gradually the pups begin to cluster, and when they are about ten days old they sleep together in densely packed pup pods. These are, are a rule, on neutral ground between the territories of the males. Here the pups are rarely disturbed by trampling bulls, they do not have to waste energy in fleeing, and heat loss is minimized by lying close together.

Four or five days after giving birth, the females go to sea to feed. Females come ashore singly, but they nearly always leave the beach en masse. Slowly and surreptitiously, ten to twenty females ease seaward. Then, suddenly, as if on command, they all dash into the sea, thus thwarting the designs of the beach bulls. With so many rushing females to choose from and so many competitors to fight, the beach bulls become thoroughly flustered and only rarely do they succeed in detaining a female.

When a female returns, she goes to the general area of the beach where she last saw her pup and begins to call it, emitting one stentorian call about every thirty seconds. Her pup, ensconced in a pup pod, wakes, answers, scrambles eagerly across its resting fellows, and rushes towards its mother. They sniff to confirm identification, she reclines, and the pup begins to nurse. The system is very efficient. On a beach crowded with roughly a thousand sea lions, Cawthorn found that the average elapsed time between a returning female's first call and reunion with her pup was eleven minutes.

Initial feeding trips are brief. The females return, as a rule, in twelve to fifteen hours. Later they stay away for several days, and in the absence of the females, the pups get hungry. Many begin to wander about the rookery, attempting to filch milk from any handy cow. They usually pick on one that is feeding her pup, sidle up quietly, and nurse at one of the other teats. The cow awakes and snarls, and the freeloader backs off, but soon it nurses again. The risk is minimal because the placid females threaten but never harm these moochers. I once watched a female suckle four pups.

When they are about one month old, the pups begin to disperse. They move onto the sward, play with each other, and are often mercilessly pestered by the immature males, which try to assemble pup "harems" and occasionally attempt to copulate with pups. Partly to escape these ruffians, partly out of curiosity, and also because the feeling of being enclosed and touched seems to appeal to them, many pups at this time crawl into rabbit burrows to their deaths.

By mid-January all females at Sandy Bay had given birth to their pups (the first was born on December 10) and nearly all had mated. They, too, now abandoned the beach and moved inland, while the adult bulls, now lean and spent, their fur ragged and lusterless, began to leave the island.

The pups moved farther and farther inland. They no longer clustered in pods. Most slept singly, or in twos and threes, beneath the dense myrsine bushes. A few wandered far into the forest. In early February the females started to take their pups to sea. The pups, some weighing nearly 60 pounds (27 kg), had serious misgivings about this move, but the females were patient and extremely persistent. They called and coaxed, returning again and again to encourage the reluctant, lagging pups; one I watched took nearly an hour. The pup followed its calling mother obediently, but at the water's edge it rebelled. The mother swam out and returned a dozen times, while the pup, bawling disconsolately, advanced and retreated with the waves. Finally a big breaker caught it and carried it off. It struggled frantically in the roily surf, heard its mother's call, and paddled awkwardly toward her voice. She surfaced near it, nuzzled it, and turned. The pup followed, head held high and no longer panicky, and together they swam out to sea.

25

At Play in the Fields

PHYLLIS DOLHINOW

If the hours a young monkey spends each day in play are any indication, then play must be a major category of primate behavior. This conclusion is underscored by the complexity of play and by the amount of energy a young monkey devotes to it.

Play is probably important in the development of all mammals, but it appears to be particularly important for the slow-maturing monkeys and apes. Juvenile monkeys play for years, investing thousands of hours of activity, energy, and emotion. Such an expenditure of biological resources must serve important biological functions. The theory of natural selection compels us to look for the adaptive reasons for this behavior.

Perhaps in part because of the values of our culture, play has not been considered a major research problem. Field studies have relegated play to one of a long list of kinds of behavior. To appreciate the subtleties of its performance and to attempt to understand the nuances of its functions, play must be evaluated in the natural setting of the species, rather than in a laboratory.

For most primates the context of normal life is a social group—in all its complexity and stability. This group is a small world with few intrusive events. A majority of monkeys live their entire life in the group of their birth; hence they know it and its location well. This is the setting of play: a rich blend of social relationships and ecological pressures, dangers, times of plenty and scarcity, and seasonal changes in both environment and group. Play behavior is characteristic primarily of large infants and juvenile monkeys, although adults may play on rare occasions. This fits in with the notion that play is prepartation for adult life, that it is of major importance in the learning process. In contrast to adult activities, play is its own reward. Play does not lead to the attainment of some other goal, such as food. The playing juvenile uses the same kinds of behaviors as the adult, but often in odd sequences or combinations. Play fighting may be aggressive, but it still includes actions that would be suicidal in a serious encounter. Little monkeys make great efforts to play: they go to other juveniles, initiate games, and stimulate each other. In this sense, play comes from within, it can be pleasurable, and its actions are repeated over and over again, year after year.

Superficially, play may seem simple. There is chasing, wrestling, and boxing. But closer inspection shows that the actions are not simple. If it is an important part of the education of a species and if many adult behaviors are practiced in games, then we should expect to find great variety behind the apparent simplicity of play. This is, in fact, the case.

When a large number of monkeys play, there are rapid alternations of participants, actions, and individual moods, which all produce variations of activity. Some motor patterns appear regularly. A group of playing rhesus monkeys may, in a few minutes, push, pull (hair, fingers, limbs, ears, tails), spin, squint, drag feet, rub with hands, scratch with nails, run, jump, chase, fall down, charge, swing by the feet, lean, mouth, and shove. The list usually also contains gestures of threat and submission. When threats appear, tension may develop, turning play into aggression.

The intensity of play can be measured only imprecisely, The human observer cannot use the frequency of cries as an index of pain; cries may reflect the mother's closeness and willingness to back up her infant more than it reflects how much the infant is hurt. Social context can affect play in ways that are not obvious. An animal may ignore pain in play if the intentions of the inflicter are nonaggressive, whereas in other interactions a bite or slap of the same intensity will produce a severe reaction. The sound of heads banging on tree limbs or on hard ground in play makes the human observer wince, but often appears not to deter the monkey whose head was banged. Instead, he sits for a moment, then jumps back into the fray.

Subtle glances, tensing of muscles, pressure of grasp, severity of nipping are but a few of the many cues that are not available to the human observer, who is aware of only a small portion of what the players experience.

Consider the following examples of play, which I have observed among wild primates:

1. In a village in northern India, a group of four large and two small juvenile langurs were playing on top of an abandoned irrigation well. The two largest sat face to face boxing each other's shoulders. Next to and touching them was another pair of juveniles wrestling in a ball. All four

were bumped by the small juveniles who ran around and over the well in a wild game of chase and reverse chase. From time to time individuals would pause, look around them, and flop to the wall or ground for a few seconds, as though gathering strength for another onslaught of wrestling or boxing. Occasionally, partners changed; and even the smallest langur eventually sparred with the largest, although for a much briefer time than a pair equally matched. Most of the action took place in pairs, and the participation of each langur appeared to be completely voluntary.

2. In the Singapore botanical gardens, two crab-eating juvenile male macaques were sitting close to one another on a limb. The larger was looking at the smaller, who was quietly regarding his navel. The larger got up, glanced again at the smaller, reached out and poked him with a finger, then quickly sat down in the same spot. The smaller jerked his body away from the larger and looked in the opposite direction. After a second's hesitation the smaller looked toward the larger, who immediately slapped him again. The small monkey jumped four feet back along the branch and the larger immediately followed and grabbed the retreating animal's tail. There was a momentary tug-of-war, which ended with the younger falling from the branch and the larger holding him by the tail in midair. When his tail was released, the smaller fell to the ground. The larger macaque then jumped down and chased the smaller one out of sight. Depending upon how this was recorded, it could be tallied as play or aggression.

3. A small juvenile rhesus monkey was roughhousing with an infant on a forest pathway in northern India. The infant was fairly passive and the juvenile turned him over and over quite roughly, mouthing him all the while. The infant began to make faint "uh-uh-uh" sounds, and the juvenile immediately paused in his handling, glanced around to see which animals were nearby, and them resumed mauling the infant. The infant used the vocalization repeatedly whenever the juvenile appeared to be rougher than the infant could tolerate. Always the response of the juvenile was to cease, look and check the reactions of all nearby adults, and when there were none, to continue.

At one point the juvenile pushed the infant forcefully against a limb and the thud of the impact could be heard for 20 feet. The infant squealed sharply and the juvenile stopped. An adult female sitting nearby moved toward the pair and the juvenile ran off. In this instance the infant was using the vocalization very skillfully to force the juvenile to modulate his play activity. The infant was, in fact, controlling the "play" situation. Whether the activity was in any way pleasurable for the small monkey is questionable; that it was for the larger, seemed apparent.

4. Two small rhesus monkeys were sitting side by side on a rooftop

in Lucknow, India, and one reached out and leaned on the other. The leaned-on one moved three feet away and sat down. The first again leaned toward him, this time reaching out and cuffing him lightly on the knee. After the slap the first rhesus bent back to a normal sitting position and looked solemnly at the little monkey he had just slapped. The latter sat quietly and gazed directly ahead, away from the other monkey. The first rhesus repeated the slap, this time with a little more force; still no response, then another slap—each time he leaned forward and then quickly bent back. Finally, he reached out, grabbed a handful of fur, and tugged at it hard enough to pull the skin in a fold. The solicited monkey grimaced at him and bent as far away from him as he could without standing and moving. The rhesus that had been trying to gain the other's attention then sat quietly and after a minute moved away.

In another, similar instance, where the two monkeys were juveniles, the one who was slapped responded after several approaches with a threat and attempted to bite the slapper. This started a fight that was broken up by a dominant adult male. In a a third example, one young monkey tried to solicit play in a similar manner, but in this instance the response of the slapped monkey was to join the first in boisterous wrestling and chasing play for some time.

A special gesture called a "play face" is described for many species and signals the nonaggressive intent of the monkey or ape that wants to play. This is an oversimplification, since the monkey soliciting a play partner signals in many ways. It may roll its head from side to side, close its eyes, move in an uncoordinated, jerky manner, bow or bob up and down, or approach backward. The face is only one element of a complex set of movements that carry the message of intent to interact in a manner we call playful. An invitation to play might be a slap and quick retreat; it might even be a bite or a shove, repeated in different ways. There are many ways to start an interaction, to determine the mood of the desired partner, and to communicate the intention of the solicitor.

Field workers who have observed monkey infants grow to maturity can suggest many benefits that derive from social play, including the practice of a number of social gestures and motor skills. Patterns appear in fragments, hardly recognizable as the stereotyped social signals they will become if they are to be effective signals among group members. These patterns are practiced over and over, in myriad contexts and among all the young of the troop. There is a slow but certain increase in the motor skills of each monkey, from the time it first leaves its mother, stumbling off a few inches to investigate bits of its environment, to the time when it will leave her to play for hours. Although it is unlikely that we would call the very early sensory or motor experimentation play, there is no question about much of the later activity of the older infant.

Motor and social skills are practiced in play, but it has not been demonstrated that play is the context in which these skills are originally stimulated or learned. This distinction must be made if the benefits of playful activity are to be understood. If play is considered a context for consolidating skills, for adding small increments of ability and mastery over motor, manipulatory, and social tasks, then the vast repetition of play patterns makes a great deal of sense. Play is not for solving problems from scratch, unless perhaps they are problems with objects, but it may help once the process of solution has been started.

Repetition is a key descriptive word for play—and repetition is essentially practice. Elements of sexual behavior and dominance gestures appear early in social play, but they appear as fragments and often in no apparent relationship to reproductive or actual dominance contexts until the monkey is much older. Playing animals are involved in a great deal of physical contact and continuous social interaction. The ability to control one's own behavior and the actions of other monkeys becomes very important.

Given the amount of direct interaction—especially physical contact—in play, each monkey soon learns differences in the size, strength, reaction time, and tolerance of each player. Rules of dominance are essentially based on strength and ability to use social signals, and if learning which animals are stronger involves some pain, the young monkey may learn the rules rapidly. For young juveniles, especially males, the opportunity to play dominant, as well as subordinate, roles may be a part of the attractiveness of play. From a broader view, the total experience of play makes ranking possible and seemingly inevitable.

Social cues and complex communication patterns are developed in the relative safety of play. It does not do any good to be the strangest and largest in the group if at the same time most other adults can bluff their way past to a desired object. A monkey must know not only the form and context of each social gesture, it must also be able to execute each with style and finesse. Timing must be perfect, and since most fights are avoided by complex gestures of threat and submission, the monkey that bluffs best probably goes furthest in the long run.

Play is often considered by humans to be pleasurable. Among free-ranging monkeys some play appears to be pleasurable and fun, but much does not. The tensions aroused during play often appear to result in the dissolution of the play group. It is possible that a degree of what we think of as pleasure is obtained from increased competence of movement and skill in the use of signals. We can only guess the motivations of a nonhuman primate; it cannot tell us anything from introspection.

A juvenile male wrestling in the arms of an adult male may appear tense and inhibited in his movements. He may finally utter a squeal of fear and succeed in breaking free from the adult. The latter may have been making a play face toward the younger male the entire time they were in contact, but the gesture did not avert the juvenile's breaking away. What may surprise the observer is the juvenile's immediate return to the adult male; this pattern may be repeated again and again, with the juvenile fleeing each time, only to return for more rough play. The juvenile's ambivalence is apparent; his actions continually shift from approach to withdrawal. Such conflict situations, with their tensions and anxieties, are present in the behavior of monkeys and apes. It remains to be demonstrated whether or not any of them are resolved in play.

An ape may pound on a tree when it cannot pound on another ape that annoyed or frightened it. Whether a young ape destroys a twig in play because it would like to do the same to a new sibling is a matter of human conjecture. The play group may be a location for working out aggression that the animal might otherwise wish to direct toward larger and stronger individuals. Helplessness for the younger infant monkey relates most clearly to locomotion and anxiety about leaving its mother. As it grows, it will also experience anxiety related to relationships with others.

Much play appears to be testing of one kind or another. Players constantly push to the limits of tolerance of aggressive behavior, especially older male juveniles that are large and strong and able to inflict injury. An aggressive invitation can be followed by avoidance, play, or fighting—depending upon the mood of the solicited, the actions of the solicitor, their past experience, nearby animals, and probably a long list of other factors that are not apparent to the human observer. No wonder the human reports that rough play borders on serious aggression and that it is difficult to know whether to fit these episodes into aggression or play. The apparently ambiguous actions of large immature monkeys doubtless reflect their ambivalent feelings as to whether they will fight, mildly test relative rankings, or tussle and chase in a frankly playful manner.

The immature monkey constantly tests its strength and social skills, its bluffs, evasive abilities, and allies in playful activity. It also tests its environment, but not, at first, in play. Strange surroundings or unusual events appear to inhibit playful activity. New corners of the environment are first investigated; then, and only then, played in. The initial response to strange objects or events is one of great caution.

As a way of testing and gaining small increments of skill or mastery, play is tremendously important for the individual. Play has been described as uneconomical, but this is a judgment that ignores its long-term benefits. In terms of immediate goals, playful activity is expensive in energy and time, but if the eventual behaviors of the adult are considered, it is a good investment of both.

Play is one of the most important factors in the establishment of social relationships that last a lifetime. The nonhuman primate is born into the highly structured social context of the group and the specific relationships its mother has with the group. Her personality directly and indirectly affects the infant's contacts with other group members—the effect may be restricting or it may be encouraging of wide contact. If the mother is very subordinate and constantly tense when she is near other adults, she may stay away from most of the adults and deliberately restrict the movements of her infant so that it will not be able to play. If, on the other hand, she is a confident, dominant, and socially active female, her infant may be in the center of action and have a lot of contact with other monkeys. A mother that is quick to threaten young that solicit her infant to play reduces the total amount of time her infant spends in play. Her presence, regardless of her temperament, certainly facilitates early exploration and play. Young of dominant females can afford to take liberties against other monkeys when the mother will back them up against reprisals.

Various types of play characterize different stages of development, each with its attendant problems and challenges. Play and other activities must have been designed through evolution to help meet the demands of each level of maturation. While we can observe social and motor tasks being worked out by infants and juveniles, the human can only guess at the psychological tasks. We cannot determine what, if any, utilitarian internal functions play has for the monkey.

Playful behavior gradually drops out of the repertoire of the monkey as it matures. Most adult male monkeys rarely, if ever, play, and the situation is similar for adult females although they may interact playfully with their infants. In any event, play is not a notable feature of adult life. Why does it drop out along the way and why don't adults play?

There is no satisfactory or final answer to these questions, but it does seem that the following factors are relevant. The adult may find it too difficult to indicate playful intentions. The signals of playfulness, including the so-called play face, may not be sufficient to counteract the strength and potential ability to damage that other members of the group have learned to associate with that adult. Most play involves a lot of physical contact and sudden movements—two qualities of interaction generally avoided by adult males unless the situation is clearly one of relaxed grooming or similar activity. The rough and tumble of play may be potentially too dangerous for adults and also incompatible with their important roles of leadership and dominance. Intentions that are ambiguous or misread only once could mean the difference between safe play and a serious wound or fatality.

The adult monkey, and especially the adult male, is generally very sensitive to which other adult comes close to him. A sudden invasion of the adult's personal space or the area around him that he considers his private domain might be disastrous. The normal tensions of dominance relationships are seldom evident; in part they are hidden because the actors in the structure carefully avoid getting into situations where positions must be challenged. One of the best ways to avoid a fight is to avoid physical proximity. Play is thus to be avoided.

Aside from these considerations—of animals getting too close, invading each other's personal space, or misreading the signals of intention—there is another, more basic suggestion as to why it is not worth the risk for adults to play. The learning activities of play, so important for the infant and juvenile, are no longer necessary for the adult. Presumably, by the time the individual has matured, it has mastered the skills it will need, learned the land it occupies, established its social relationships, and become coordinated motorically. Major forms of adult behavior are established and relatively immutable.

Mammals are so constituted that learning takes incredible repetition. Mastery comes slowly, and the years of immaturity are the time of life when they can afford the most mistakes. Increments of skill can be seen only over time and are based on the repetition and practice of activities like play.

Laboratory studies have demonstrated the importance of peer contact for the young monkey, but only field studies have indicated the full complexities of social life and the environment and their importance to the developing primate. The dangers and challenges of life in nature and the rewards to the individual of successful social life are only apparent in the field. The multiple functions of play may be obscure, indeed, when viewed for only a short time in an artificial context, and are not even obvious from watching only the young. The juvenile patas monkey jumping up and down in the tall grass may seem to be enjoying a nice sport, but the full importance of these motor patterns is not obvious until you see an adult male patas jumping to divert the attention of a lion from the rest of the group. Then the significance of this life-saving skill will reach the observer.

The healthy young monkey plays. It does so for a substantial portion of its immature years, and to a significant degree the success of its adult life may depend upon the intensity and variety of its play experience. Play in monkeys is more complex than the word signifies to most humans. There may be no fun in play, and it might be tension and anxiety producing for the playing monkey. Whatever the differences in form that play takes among the many different species of nonhuman primates, it is a major category of adaptive behavior that must be analyzed if we are to understand primate behavior.

Index

Acorn woodpecker: cooperative breeding, 139
Acoustic sensillum: moths, 35
Adélie penguin: nests, 91
Aggression: play of young monkeys, 234
Allepeira web, 104
Alligators: maternal behavior, 171
Altricial mammals, 122–23, 125
Altruistic behavior: Belding's ground squirrels, 135–36
Amaurobius web, 100
Amboseli National Park Baboon Project, Kenya, 205–14
American bald eagle: nests, 92
Ampullae of Lorenzini, 29–30, 33
Anchovies: schooling behavior, 164, 166
Animal behavior: development of, 11; evolution of, 11–12; function of, 12; immediate causation of, 10
Animal behavior, science of, 1; development of, 3; function of, 2–3; levels of integration in, 7–9
Animal conservation, 2–3
Anna's hummingbird: territorial behavior, 148–54
Anoles: tongue flicking, 175
Anthropomorphism, 3–7
Antipyretics, 106, 108
Antlion larvae: pit construction and predatory behavior, 44–54
Ants: slave-making behavior, 155–61; temporal polyethism, 195
Apes: play, 234
Arachnids: silk glands, 98
Araneus diadematus web, 53
Arctic tern: migration, 55
Arctiid moths: sense of hearing, 43
Argiope web, 104
Army ants, 9–12, 159
Auckland Islands, 221–22

Auditory sense, *see* Hearing
Avoidance-conditioning, 69

Babblers: cooperative breeding, 139
Baboons: mother-infant relationship, 203–14
Bacterial infections: fever, 109–11
Bald eagle: nests, 92
Bald-faced hornet: directional movement, 23
Banded geckos: vomeronasal system, 171
Bank swallow: nests, 90
Banner-tailed kangaroo rat: mounds, 215–20
Bats: echo location system, moths' response to, 34–35, 38–41, 43
Bee-eaters: cooperative breeding, 139
Bees: directional movement, 20, 26
Beetles: directional movement, 24
Belding's ground squirrel: social behavior, 129–37
Biological control of pests, 24
Birds: febrile response, 111; migration, 55–56; navigation, 55–65; nest building, 87–90; plumage differences, 113
Bird's-nest soup, 93
Black, buffalo weaver: nests, 95
Black-tailed prairie dogs: coteries, 216–17
Blotched blue-tongued skink: tongue flicking, 174
Bluebirds: migration, 55
Bluejays: toxic insect larvae, response to, 85
Body temperature, regulation of, 107–8; hummingbirds, 147; *see also* Fever
Brain: octopus, learning in, 79–81
Broad-headed skink: maternal behavior, 171
Brown, Jerran L., 145, 217
Brown lemming: migration, 190
Buffalo weaver: nests, 94–95
Bulb flies: directional movement, 25
Bullhead: electrical sensitivity, 28–29; hearing 67–68

Bull shark: hearing, 68
Bush babies: maternal behavior, 217; reproductive strategies, 124–25
Butterflies: cardiac glycosides, 85; directional movement, 25–26

Cabbage buttterfly: directional movement, 24–25
Calamistrum, 98
Caliology, 87
Callows, 11, 195
Canada goose: nests, 91
Carp: schooling behavior, 165
Carrier pigeon, *see* Homing pigeon
Cassin's malimbe: nests, 94–95
Catfish: electrical sensitivity, 28–29; hearing, 66–67, 73
Cats: colorblindness, 67
Cave-nesting birds, 93
Cavity-nesting birds, 89–91
Cawthorn, Martin, 221, 223, 226–27
Celaenia web, 105
Chacoan peccary, 177, 180
Checkered whiptail lizard: tongue flicking, 174
Chimpanzees: hunting behavior, 1–2
Chuckwalla: vomeronasal system, 171–72
Cladomelea web, 105
Classical conditioning, 69
Cliff-nesting birds, 93
Cliff swallow: nests, 94–95
Coalfish: schooling behavior, 166
Coclotes web, 102
Cold-blooded animals, 109–10
Collared lemming: endocrine system, 192; migration, 190
Collared peccary: social behavior, 177–85
Coloration, significance of: Anna's hummingbird, 151–53; ground-nesting birds, 92; snow geese, 113–19
Common blue-tongued skink: tongue flicking, 174
Communication: Anna's hummingbird, 148, 151; kangaroo rat, 215
Competition: Belding's ground squirrels, 134–35
Compound bird nests, 95
Conditioning, 69
Conservation of animals, *see* Animal conservation

Convergence: nest forms and structures, 88
Cooch, Graham, 113, 115–16, 119
Cooperation: Belding's ground squirrels, 135–36; peccaries, 180, 182–85
Cooperative breeding: Florida scrub jays, 139–46
Coraciiform birds: nests, 89
Coteries, prairie dog, 216–17
Cotinga: nests, 90
Courtship: lizards, 171
Cribellate spiders: webs, 97–98, 100–101
Cribellum, 97–98
Cypriniformes: hearing, 67
Cytophora web, 104

Darwin, Charles, 5, 115, 159
Decibel, 68
Deer: home range, 4
Defensive behavior: Anna's hummingbird, 148–52
Descent of Man, The (Darwin), 5
Desert iguana: febrile response, 110–11; vomeronasal system, 169–70
Dicrostichus web, 105
Dijkgraaf, Sven, 28–30, 73
Dinopidae web, 101
Diplellate spiders: webs, 97–98, 101–5
Dipluridae web, 102
Directional movements: birds, 55–65; insects, 19–27; *Polyergus* ants, 156–57; sharks, 33
Dispersal patterns: Belding's ground squirrels, 135; kangaroo rats, 217–20
Dogfish shark: electrical sensitivity, 32–33
Dominance: female baboons, 211–12; monkeys, 233
Doodlebug larvae, *see* Antlion larvae
Doodling: antlion larvae, 45, 47–48
Dorsal gland: peccaries, 179–80
Dragonfly: directionial movement, 23
Drone fly: directional movement, 25–26
Ducklings: parental recognition, 197–202

Eagles: nests, 92–93
Ecology: characteristics of birds' nests related to, 88–89
Ecribellate spiders: webs, 97–98, 101–5
Ectothermic vertebrates, *see* Cold-blooded animals
Electric fish, 29
Emlen, Steven, 55–57

Emperor penguin: nests, 91
Endangered species, 126
Endocrine system: lemmings, 192
Endogenous pyrogen, 108, 111
Endothermic vertebrates, *see* Warm-blooded animals
Episinus web, 103
Eresus web, 100
Estrous cycle, evolution of, 122
European nuthatch: nests, 90
European starling: nests, 90
European stone curlew: coloration, 92
European warbler: navigation, 57
European white stork: nests, 92
Euryopis, 103
Eutheria, 120
Evolution, 5, 8; behavior, 85–86; fever, 110–12; nest building, 87–96; reproductive strategies, 120–26; spider webs, 98–105

False scorpions: silk glands, 98
Fever, 106–12
Fever therapy, 107–9
Field vole: breeding, 187
Filistata web, 100
Finches: nests, 94
Fish: electrical sensitivity, 28–29; hearing, 66–74; schooling behavior, 162–67
Flickers: nests, 90
Flies: directional movement, 23–26
Florida scrub jay: social system, 138–46
Following-response: precocial birds, 197
Food seeking: vomeronasal system of lizards, 170
Food sharing: peccaries, 182–83
Food supply: defensive behavior of Anna's hummingbirds, 148–53; Belding's ground squirrels, 133–34; reproductive strategies related to, 123–24
Formica ants: slavery, 156–60
Frisch, Karl von, 20, 66
Frontinella web, 103
Fruit flies: perception of polarized light, 20

Galapagos finches: nests, 94
Galileo Galilei, 107
Gasteracanthine spiders: webs, 104
Geese, *see* Canada goose; Snow goose
Geolycosa burrow, 101

Gila monster: vomeronasal system, 169–70
Goldfish: hearing, 72; schooling behavior, 165
Gorget display: Anna's hummingbird, 148, 151–52
Gray-breasted jay, *see* Mexican jay
Green iguana: vomeronasal system, 172
Gregarious mammals, 216–17
Grey-headed social weaver: nests, 95
Grizzly bears: Yellowstone National Park, 2–3
Ground-nesting birds, 91–92; ducks, 198, 202
Ground squirrels: social behavior, 129–37

Habrobracon: predatory and sexual behavior, 17
Hearing: duckling-parent bonding, 198–202; fish, 66–74, 163–64; moths, 34–43; peccaries, 180
Heavy water, 62
Herring: schooling behavior, 162, 166
Hexura web, 102
Hibernation: Belding's ground squirrels, 131–33
Hippasa burrows, 101
Hippocrates, 106
Hole-nesting ducks, 198, 202
Home range: deer, 4; peccaries, 178–79
Homing pigeon: navigation, 58–65
Honeybees: directional movement, 20
Hooker's sea lion: behavioral development, 221–28
Hornets: directional movement, 23–24
House wren: nests, 89–90
Hummingbirds: nests, 88, 93; territorial behavior, 148–54
Hyperthermia, 108
Hypothalamus, 107–8
Hyptiotes web, 100

Icterids: nests, 94
Iguanas, *see* Desert iguana; Green iguana; Iguanid lizards
Iguanid lizards: tongue flicking, 174–75
Imprinting, 196; precocial birds, 197, 201; snow geese, 116–18
Independent colony founding: ants, 160
Indigo bunting: navigation, 56
Infanticide: Belding's ground squirrels, 134
Infections, fever caused by, 107–12
Inner ear: fish, 66–67, 72, 74

Insects: directional movements, 19–27
Intelligence, 5–6
Internal clock: pigeons, 60–62

Jacamara: nests, 90
Jacks: schooling behavior, 164
Jacobson's organs, *see* Vomeronasal system
Jays, *see* Bluejays; Florida scrub jay; Mexican jay

Kangaroo rats, *see* Banner-tailed kangaroo rat; Merriam's kangaroo rat
Kidnapping: infant baboons, 208–9
Kingfisher: nests, 89–90
Kinship and social behavior: Belding's ground squirrels, 135–36
Kiwi: nests, 89
Knifefish: electric field, 29

Ladybird beetle: directional movement, 24
Langur: play, 230–31
Lateral line system, 66–67, 73–74; and schooling behavior, 164
Learning: octopuses, 75–83; play of young monkeys, 232–33, 235–36
Lemmings, *see* Brown lemming; Collared lemming; Norway lemming
Lemur: reproductive strategies, 124–25
Life span: Belding's ground squirrels, 133; Norway lemmings, 191
Linyphiidae: web, 103
Liphistius: web evolution, 98, 101–2
Lizards: febrile response, 110–11; tongue flicking, 168–76
Lorenz, Konrad, 118, 196–97, 201
Lorquin's admiral butterfly: directional movement, 25
Lycosid spiders: silk, 100–101

Macaque: play, 231
McCook, H. C., 104
Mackerels: schooling behavior, 163–65
Magnetic field: effect on homing pigeon navigation, 63–65
Malaria, 106
Male-infant relationship: baboons, 208–10
Mallards: duckling-parent bond, 198–201; plumage differences, 113
Mallee fowl: nest mounds, 89

Mammals: febrile response, 111–12; reproduction patterns, evolution of, 120–26
Map and compass system of bird navigation, 56–57
Mastophora web, 105
Maternal behavior: baboons, 203–14; banner-tailed kangaroo rats, 219; ducks, 199–201; monkeys, 235; peccaries, 183–84; reptiles, 171; sea lions, 225–28
Mating; peccaries, 184; snow geese, 114–19; *see also* Courtship; Reproduction strategies; Sex recognition
Megapodes: nests, 88–89
Merriam's kangaroo rat, 217
Metatheria, 120
Metepeira web, 104
Mexican jay: cooperative breeding, 145–46; dispersal, 217
Miagrammopes web, 101
Microbar, 68
Microlinyphia web, 103
Migration: birds, 55–56, 65; Norway lemmings, 186–91
Minnows: hearing, 67
Mites: silk glands, 98
Monarch butterfly: cardiac glycosides, 85
Monkeys: play, 229–37
Monk parakeet: nests, 95
Monogamy: Florida scrub jay, 141; snow goose, 117
Morell, Benjamin, 222
Mosquitoes: directional movement, 24
Moths: color polymorphism, 119; evasion of moth-eating bats, 34–43
Motion sickness, 85
Mound-nesting birds, 88–89
Mourning dove: nests, 92
Muscid flies: directional movement, 24

Natural selection: color polymorphism, 119; reproductive strategies, 122–23
Navigation, *see* Directional movements
Neoscona web, 104
Nephila web, 104
Nepotism: Belding's ground squirrels, 135–36
Nervous system: soft-bodied invertebrates, 80
Nest building: evolution, 87–96; primates, 124–25
Neuromasts, 164

Noctuid moths: hearing: 35–36
Norway lemming: mass migration, 186–93
Nursing behavior: peccaries, 183–84; sea lions, 225–27
Nuthatch: nests, 90
Nutritional immunity, 112

Obligatory parasite, 160
Octopuses, learning in, 75–83
Olfactory sense, *see* Smell, sense of
On the Origin of Species (Darwin), 5
Osprey: nests, 92
Overpopulation, *see* Population explosion
Ovulation, evolution of, 122–23
Owlet moth: hearing, 35

Pachygnatha web, 104
Painted lady: directional movement, 24
Painted snipe: nests, 91
Palm chat: nests, 95
Parakeets: nests, 95
Parasitic wasps: directional movement, 24; predatory and sexual behavior, 17
Parasitism, 155; social parasitism of slavemaking ants, 155–61
Pardosa burrow, 101
Parental recognition: ducklings, 197–202; sea lions, 225–27
Parker, G. H., 66
Parrots: nests, 89–90
Passerine birds: nests, 93, 95
Pavlovian conditioning, *see* Classical conditioning
Peccaries, *see* Chacoan peccary; Collared peccary; White-lipped peccary
Penguins: nests, 91
Peppered moth: color polymorphism, 119
Pheromones: army ants, 10–11; *Polyergus* ants, 157; lizards, 173; schooling behavior in fish, 163
Philopatry, 118
Pholeus web, 102
Phylogenetic scale, 125
Piciform birds: nests, 89
Pigeon, *see* Homing pigeon
Pigs, 177–78
Pirata nest, 101
Pityohyphantes web, 103
Play, 196; baboons, 204–5; monkeys, 229–37;

peccaries, 182
Play face, 232, 234
Poisonous foods, protection against, 85
Polarized light, 20–22; directional movements of insects, 19–27
Polyergus ants: slave-making behavior, 155–61
Polymorphism: color phases in snow geese, 113–19
Poor-will: body temperature: 88
Population crash: Norway lemmings, 191–92
Population explosion: Norway lemmings, 188–89, 191
Positive assortative mating: snow geese, 114–19
Prairie dogs: coteries, 216–17
Precocial birds: following-response and imprinting, 197, 201
Precocial mammals: reproduction strategies, 121–26
Predation, 155; Belding's ground squirrels, 134–36; parasitic wasps, 17; schools of fish, 166–67; vomeronasal system of lizards, 172
Primates, evolution of reproduction patterns in, 121–26
Prolinyphia web, 103
Propaganda pheromone: *Polyergus* ants, 157
Prostaglandins, 108
Protective coloration: ground-nesting birds, 92
Prothonotary warbler: nests, 90
Prototheria, 120
Psechridae webs, 100
Ptarmigan: coloration, 92
Puffbird: nests, 90
Purse web spider, 102
Pythons: maternal behavior, 171

Rabbits: Enderby Island, 221
Racing pigeon, *see* Homing pigeon
Ray spider: web, 104–5
Redcockaded woodpecker: cooperative breeding, 139
Red-leg infection, 110
Reproduction strategies: Belding's ground squirrels, 131–32, 136; evolution of, 120–26; parasitic wasps, 17
Reptiles: febrile response, 111
Rhesus monkey: play, 230–32
Rieffer's hummingbird: nests, 88
Robins: migration, 55; nests, 92–93

Roofed bird nests, 93–95
Rough-winged swallow: nests, 90

Saint Vincent tree anole: tongue flicking, 175
Sandpiper: nests, 91
Scent marking: peccaries, 179–80
Screw worm fly, 2
Sea lion, *see* Hooker's sea lion
Sensory adaptation, 37
Sex recognition: lizards, 171–73
Sharks: electrical sensitivity, 28–33; hearing, 67–68, 73
Side-blotched lizard: vomeronasal system, 171
Sight, *see* Vision
Silk, spider, 97–99
Silversides: schooling behavior, 162, 164
Sisyphus, legend of, 51
Skinks: maternal behavior, 171; tongue flicking, 174; vomeronasal system, 175
Slavemaking ants, 155–61
Smell, sense of: fish, 163–64
Snakes: tongue flicking, 168
Snow goose: color polymorphism and mating behavior, 113–19
Sociable weaver: nests: 95
Social behavior, 127–28; kangaroo rats, 216, 219–20; peccaries, 178–85; play of young monkeys, 229–37
Sociobiology, 128
Sockeye salmon: migration, 5–6
Solitary mammals, 217
Sooty tern: nests, 92, 95
Sparrow: nests, 91
Species recognition: lizards, 171–73, precocial birds, 197, 201
Spectacled weaver: nests, 94
Spider webs, 53; evolution, 97–105
Spinnerets, spider, 97
Starling: nests, 90
Stars, navigation of birds by, 55–57
Stegodyphus web, 100
Stork: nests, 92–93
Sun, navigation of homing pigeons by, 60–62
Sunbird: nests, 94
Sunfish: schooling behavior, 165
Swallows: nests, 88, 90, 93–95
Sweating, 107–8
Swiftlet: nests, 93
Swifts: nests, 93

Swim bladder, 66–67, 72–74
Sybota web, 100, 104

Tarantula web, 101–2
Taste, sense of: octopuses, 77
Taxonomy, 8
Teleutomyrmex schneideri, 160
Temporal polyethism, 11, 195
Termite nests, birds that breed in, 90
Territorial behavior: Anna's hummingbirds, 148–53; Belding's ground squirrels, 134–35; Florida scrub jays, 140–46; kangaroo rats, 217; lizards, 172, 174–75; peccaries, 179; sea lions, 223–25; social parasitism and, 159
Territorial budding: Florida scrub jays, 143–44
Tetragnathid spiders: webs, 104
Theridiid spiders: webs, 102–3
Thermometer, 107
Titmice: nests, 94
Tongue flicking: lizards, 168–76
Touch, sense of: octopuses, 76–79
Trap-door spiders: web and burrow, 98, 101
Tree lizard: tongue flicking, 174
Tree-nesting birds, 92
Trogon: nests, 89–90
Tropical birds: nests, 93–95
Tuna: schooling behavior, 164, 166
Tympanic organs: moths, 35–41, 43
Tyson, Edward, 177, 179

Ulesanis pukeiwa web, 103
Uloborus web, 100, 103

Vision: cats, 67; insects, 19–27; octopuses, 81–82; peccaries, 180; schooling behavior in fish and, 163, 165
Vomeronasal system: lizards, 168–72, 175–76; snakes, 168

Warblers: nests, 90
Warm-blooded animals, 109–10
Wasps: directional movement, 24, 26; predatory and sexual behavior, 17
Weaverbirds: nests, 94–95
Weberian apparatus, fish, 73
Whip-poor-will: coloration, 92
Whip scorpion: silk glands, 98

Whiptail lizards: tongue flicking, 174–75
Whiteheaded buffalo weaver: nests, 94
White-lipped peccary, 177, 180
White stork: nests, 92–93
Wolf spider: burrows, 101
Wood duck: duckling-parent bond, 198–201
Woodpeckers: cooperative breeding, 139; nests, 89–90

Wrens: cooperative breeding, 139; nests, 89–90

Yarrow's spiny lizard: vomeronasal system, 170–75
Yellow-rumped cacique: nests, 94
Yellow-shafted flicker: nests, 90

Zygiella web, 104